Recent Trends in
Rapid Diagnosis of Plant Pathogens

Recent Trends in
Rapid Diagnosis of Plant Pathogens

Dr. M. Reddi Kumar

Dr. B. Sreenivasulu

Dr. B. Chenchu Reddy

Dept. of Plant Pathology
S.V. Agricultural College, Tirupati,
Acharya N.G. Ranga Agricultural University.

BSP BS Publications
A unit of BSP Books Pvt., Ltd.

4-4-309/316, Giriraj Lane, Sultan Bazar,
Hyderabad - 500 095
Phone : 040 - 23445605, 23445688

Published by :

BSP BS Publications

A unit of BSP Books Pvt., Ltd.

4-4-309/316, Giriraj Lane, Sultan Bazar,
Hyderabad - 500 095
Phone : 040 - 23445605, 23445688
e-mail : info@bspbooks.net

ISBN : 978-93-52300-31-0 (HB)

Acharya N.G.Ranga Agricultural University
Rajendranagar, Hyderabad-500 030.
Phone : Off : 010-24015035
Res : 040-27425305
Grams : 'Agrivarsity'
Fax : 040-24015031
E-mail : angrau_vc@yahoo.com

S. Raghu Vardhan Reddy
Vice-Chancellor

Foreword

It is evident that one of the major reasons for low productivity is the losses caused by plant diseases. The disease problem in agriculture is as old as agriculture itself. Plant diseases are caused by different groups of organisms belonging to fungi, bacteria, viruses, phytoplasmas etc. They cause drastic yield losses in agricultural crops and therefore, need to be identified in the early stages and thereby, management strategies can be adopted accordingly.

An effective and applicable disease management strategy requires accurate diagnosis and understanding about the life and disease cycle of etiological agents. Proper and accurate identification of etiological agent and effect of environmental factors on host physiology play a very important role in designing any disease management strategy. Knowledge about detecting the pathogen before appearance of disease symptoms helps in eradication of the source of inoculum or forecasting of the disease incidence. For detection of very closely related pathogens, sensitivity of techniques is of paramount importance. The advent of molecular biology has caused a significant shift in the types of approaches used to characterize and identify plant pathogens and to devise disease management strategies. The diversity of large populations can be assessed in a relatively efficient manner using techniques like rep-PCR, AFLP and AP-PCR.

In this context, the book entitled as **"Recent Trends in Rapid Diagnosis of Plant Pathogens"** authored by Dr. M. Reddi Kumar, Assistant Professor, Department of Plant Pathology, S.V. Agricultural College, Tirupati, Dr. B. Sreenivasulu, Principal Scientist, ARS, Ambajipeta and Dr. B. Chenchu Reddy, Associate Dean, S.V. Agricultural College, Tirupathi will provide very useful information to students of undergraduate, postgraduate, research scholars, and teachers. I appreciate the efforts made by them in collecting the information and shape it into book form.

I am sure, this book would serve as a useful reference to the students, teachers, and researchers in detection and documentation of plant pathogens efficiently.

(S. RAGHU VARDHAN REDDY)

Preface

The ability to identify an organism is fundamental to all aspects of fungal diagnostics and epidemiology whether this is in the field of plant pathology, medical science, environmental studies or biological control. Conventional methods have often relied on identification of disease symptom, isolation and culturing of organisms, and laboratory identification by morphology and biochemical tests. These methods rarely allow diagnosis at the sub specific level and increasingly there is a need to identify the role of individual isolates or strains in an interaction, which is possible using molecular techniques. There is an increasing move towards molecular diagnostics of fungi, bacteria and viruses in all fields. Since its introduction in the mid 1980's PCR has provided many molecular diagnostic tools, and is readily applied to the monitoring and identification of fungi from environmental samples.

The advent of molecular biology has caused a significant shift in the types of approaches used to characterize and identify plant pathogens and to devise disease management strategies. This shift is driven by both technology and ecology and has occurred in parallel with significant changes in agricultural production methods.

The diversity of large populations can be assessed in a relatively efficient manner using rep-PCR, AFLP and AP-PCR, RAPD based genomic finger printing methods, especially when combined with computer assisted pattern analysis. Genetic diversity maps provide a frame work to understand the taxonomy, population structure, and dynamics of pathogens which provide a high resolution frame work to revise sensitive, specific, and rapid methods for pathogen detection and plant disease diagnosis. A variety of PCR-based finger printing protocols such as rDNA-based PCR, ITS-PCR, ARDRA, T-RFLPs, and tRNA – PCR have been devised and numerous innovative approaches using specific primers have been adopted to enhance both the detection and identification of plant pathogens in general and phytobacteria in particular.

In this book, we have focused majorly on some of fundamental aspects like isolation and identification of different fungal genera with their morphological characters and microscopic features, methods of isolation of nematophagous fungi and various preservative methods. Role of tissue culture in developing disease resistance plants and development of transgenics against viral and fungal diseases has also been discussed. So far as Virology is concerned, different transmission methods through which viruses contract from diseased to healthy plants, serological techniques and molecular techniques like PCR, AFLP, Multiplex PCR etc., for detection of multiple infections along with procedures are given.

Nanotechnology, as a new enabling technology, has the potential to revolutionize Agriculture and Food systems. There is an enormous interest in the synthesis of nanomaterials due to their unusual optical, chemical and electronic properties. Recent studies on the use of microorganisms in the synthesis of nanoparticles are a relatively new and exciting area of research with considerable potential for development. Application of nanotechnology in niche areas of Plant Pathology are described briefly.

In view of the nature and scope of the subject, a comprehensive book is an ideal never to be achieved. Therefore, most foreign publishers have attempted a series of textbooks covering different aspects of the subject. The students, however, often may not have access to all the volumes of the series, even if accessible, usually may not have the time to consult them all. These considerations prompted this effort to fulfil the needs of students as well as teachers of the subject.

Obviously, it is unrealistic to claim any degree of competence in all the diverse techniques covered in the book. This compilation includes information on various biological, serological and molecular techniques employed in the detection of plant pathogens viz., fungi, bacteria and viruses. Much of the information is collected from various reviews, books and practical manuals. This will be of much use to the students and scientists who are pursuing molecular biological and plant pathological studies.

We express our sincere gratitude to Dr. M. Sudarshan Reddy, Dean of Agriculture, ANGRAU, Hyderabad for his encouragement and opportunity given for the preparation of this text. Thanks are due to Dr.K.Veeranjaneyulu, University Librarian, Rajendranagar, ANGRAU, Hyderabad for his constant support in accomplishing the work. We wish the book will provide useful information to students and researchers in documentation of plant pathogens efficiently.

-Authors

Acknowledgement

The first author of the book owe a debt of gratitude to many people for their help in preparing this book. My heartfelt thanks are due to Dr. R. Samiyappan, Director, Dr. M. Ramaiah, Professor and Head, Centre for Plant Protection Studies, TNAU, Coimbatore, and Dr. V. G. Malathi Professor, Advanced Centre for Plant Virology, IARI, New Delhi for providing an opportunity to gain hands-on experience on various serological and molecular techniques at their respective centres. Services of Dr. D.K. Agarwal, Principal Scientist, ITCC, IARI, New Delhi and Dr. P. N. Chowdhry, Ex- Principal Mycologist, ITCC, IARI, New Delhi , are duly acknowledged for providing an insight in identifying different fungal genera. The financial assistance provided by APNL-BT Project on " Nematode Pests" in procuring infrastructural facilities for carrying research work is highly acknowledged.

It would be incomplete, if I fail to express my sense of gratitude to my beloved father for his eternal blessings and my wife, Madhavi and little rosy buds, Jahnavi and Charan Tej for their ceaseless support, ever lasting love and affection showered on me, who had been always driving force behind me, without whom this work would not be accomplished.

M. Reddi Kumar

Contents

Chapter 4

Nematophagous Fungi **41**

Chapter 5

Mycotoxins–Importance and their Detection **60**

Chapter 6

Isolation and Inoculation of Bacteria   75

Chapter 7

Detection of Bacterial Infection   80

Introduction

It is well recognized that crop diseases constitute a major constraint on a increasing crop production all over the world. Plant diseases are caused by different group of organisms belonging to fungi, bacteria, viruses, phytoplasmas etc. The detection of the presence and assay of pathogen population are basic requirements for the elimination of *Fons et origo* of the pathogen at the earliest possible moment to minimize the disease incidence and infection with its subsequent spread. If method is simple it is likely to be adopted in large scale provided the detection of pathogen infection can be done with reasonable accuracy. Pathogen is a very important component of disease cycle. Correct identification of pathogens is an important component of disease cycle. Because it is important to know their distribution, epidemiology and breeding crop plants for resistance. Diagnostics play an invaluable role in the formulation of managemental strategies. A brief account of the role of diagnostics in various steps of disease cycle are discussed here under.

Accurate knowledge of primary source of inoculum is very essential in the framing of any disease management strategy. Appearance of A2 mating type of *Phytophthora infestans* and global migration of various other strains of this pathogen employing molecular markers, demonstrated that strains in the Pacific Northwest USA during the late 1980s were the introductions from Northwest Mexico. PCR-based techniques when employed showed that these strains had a unique mitochondrial haplotype (Goodwin et.al., 1992). Environmental factors are very important component of disease. The environmental factor on host or the pathogen or both can be studied with the help of various molecular based techniques.

Proper and accurate identification of etiological agent and effect of environmental factors on host physiology play a very important role in designing any disease management strategy. Knowledge regarding the pathogen before appearance of disease symptoms helps in the eradication of the source of inoculum or forecasting of the disease.

Selection of healthy planting stocks is very essential in vegetatively propagated crops. Certified / healthy propagating materials can be selected through the use of sensitive and reliable diagnostic techniques.

1

Introduction of objectionable pathogens in an otherwise pathogen-free area can be prevented through proper testing of materials and in this direction modern diagnostic techniques play very important role.

Specific diagnostic tests are essential for those pathogens, which have wider host range, viz., *Erwinia, Sclerotium, Phytophthora* etc. For detection of various closely related strains of a particular pathogen such as *Erwinia* Digoxigenin-labeled DNA probes and ligase chain reaction (LCR) assays have been found to be highly useful (Ward and Deboer, 1994; Wilson et.al., 1994). Highly specific primers are used to distinguish strains of PVY in PCR assays. Serological methods are used for detectionof plant pathogens, particularly plant viruses have been available to plant pathologists for many years. However, the introduction of more advanced immuno-diagnostic methods into plant pathology has expanded the scope of application to the diagnosis of diseases caused by many different viruses, bacteria, spiroplasmas, mycoplasma like organisms (MLOs) and fungi. This book mainly emphasize the basic features of most commonly used and widely applicable immunoassays used for detection and identification of plant pathogens besides various biological methods.

For detection of very closely related pathogens, sensitivity of techniques is of paramount importance. Serology based detection of some pathogens is limited to certain months of the year and to specific parts of the plant particularly in case of some woody plants. Application of PCR detection is helpful in these situations since sensitivity of the test allows detection of the pathogen in most tissues (James et. al., 1999). Reliability of detection is further enhanced by employing IC-RT-PCR with ELISA (Jacobi et. al., 1998). Loss in sensitivity may occur when multiplex IC-RT-PCR is employed for detection of closely related viruses (Seoh et. al., 1998). It can be overcome by standardizing the detection protocols.

The advent of molecular biology has caused a significant shift in the types of approaches used to characterize and identify plant pathogen and to devise disease management strategies. These nucleic acid based techniques particularly polymerase chain reaction (PCR) have greatly facilitated genomic analyses of microorganisms and provided enhanced capability to characterize and classify strains. It also facilitated research programmes to assess the genetic diversity of populations. The diversity of large populations can be assessed in a relatively efficient manner using techniques like rep-PCR, AFLP and AP-PCR or RAPD based genomic finger printing methods especially when these are combined with computer assisted pattern analysis. The genetic diversity maps provide a framework to understand the taxonomy, population structure and dynamics of pathogens and provide a high resolution frame work to devise sensitive specific and rapid methods for pathogen detection, plant disease diagnosis, as well as management of disease risk.

A variety of PCR based finger printing protocols such as rDNA based PCR, ITS-PCR, T-RFLPs and multiplex – PCR have been devised and numerous innovative approaches using specific primers have been adopted to enhance both detection and identification of plant pathogens. PCR-based protocols, combined with computer based analysis have provided novel fundamental knowledge of the ecology and population dynamics of microorganisms. Thus, presenting exciting new opportunities for basic and applied studies in Plant Pathology.

Modern techniques discussed above have been found to be very useful in the proper understanding of basic population genetics of several plant pathogens. Analysis regarding the origin of A2 mating type of *Phytophthora infestans* indicated that the immigrant strains were more aggressive (Goodwin and Drenth, 1997; Fry, 2001). Free international trade without non-tariff barriers under WTO will be possible only when properly validated pathogen detection technologies are used (Leach, 1999).

Chapter 1

Essentials of Phyopathological Laboratory

The science of Plant Pathology is at a crossroads. A quick transformation from the use of conventional techniques in the rapid detection of plant pathogens to modern, reliable and reproducible molecular techniques like PCR has been witnessed. In this chapter, a list of equipments and instruments which are commonly used in plant pathology laboratory and the recent developments are briefly discussed.

Plant Pathologists need to study and examine structure of different microbial cells and their components which are too small to be seen by the naked eye. In fact, the study of microbes had received great attention with the discovery of light microscope which made it possible for the scientists to see primarily the different shapes of bacteria and the structures associated with them and it has become an indispensable tool for microbiologists and plant pathologists.

1.1 Microscopy

A microscope is defined as an optical instrument which consists of lens or combination of lenses making enlarged or magnified images of minute objects. Based on number of lenses, microscopes are two types viz., simple and compound microscope.

Simple microscope, which is a primitive type where magnifying glass is held in a frame. It consists of only one lens it is also called as magnifying glass. While compound microscope differ from simple microscope in having two sets of lens such as objective lens and ocular lens or eye-piece. The eye piece is mounted in a folder which is commonly called body tube.

Working Principle : It has a lens system of consisting an objective and an eye-piece. As the light passes through the condenser, penetrates through the specimen present on the stage of microscope. The objective lens produces first image, which is real, enlarged and inverted. The eye-piece further magnifies and projects the second image which is virtual and upright on the retina.

For the efficient use of microscope, one should understand the basic principles of microscopy viz., magnification, resolving power, numerical aperture and focusing.

Magnification : It is the function of a two lens system where the ocular lens is found in eye piece and the objective lens is situated in a revolving nose piece. As the light pass through the specimen, the objective lens magnifies it and produces real image which is further magnified by ocular lens and produce the final image. There are different types of objective lens containing different degrees of magnification, when these are combined with ocular lens, the total or overall linear magnification of the specimen is obtained.

Resolving Power : It is the ability of a lens to show two adjacent objects as discrete entities. Unlimited enlargement of the specimen is not possible by merely enhancing the magnification power, because the lenses are limited by a characteristic called resolving power. When a lens fail to show two objects clearly, that means lens has lost resolution.

Numerical aperture : It is characteristic of each lens. The resolving power of a lens is dependent on two important properties viz., wave length of light used and numerical aperture. It is defined as a function of the diameter of the objective lens in relation to its focal length. The relationship between resolution and NA can be expressed as $d = \lambda/2NA$

where,

d = resolution   NA = Numerical aperture

λ = Wavelength of light used

Focus and Illumination : For the purpose of illumination, artificial light from a tungsten lamp is the most commonly used light source in microscopy, since the intensity of day light is an uncontrolled variable. The light is passed through the condenser located beneath the stage. It is always keep the condenser close to the stage, especially when oil immersion objective is using. The condenser contains two lenses that are necessary to produce maximum numerical aperture. The amount of light passed will vary with objective lens used. The working distance between the objective and stage decreases as magnification of lens increased where as numerical aperture of objective lens increases.

1.1.1   Types of Microscopes – Salient features

1.1.2   Dark field Microscope

In this microscope, objects are brightly illuminated against a dark back ground. A special type of condenser allows a hollow cone of light from a source of illumination. The condenser directs the light obliquely so that light is scattered or deflected from the specimen. This diffracted light will enter the objective and reach the eye; thus the object or microbial cell will appear bright. This is similar to ordinary light microscope where viable organisms are more readily observed than with bright field microscope.

1.1.3   Bright field Microscope

The microorganisms appear dark against brightly lighted microscopic field. As microorganisms donot absorb much light, but the light absorbing ability is enhanced by staining them with a dye, resulting in greater contrast and colour differentiation, however it is difficult to observe living cells.

1.1.4   Phase-contrast Microscope

It is extremely valuable in studying the microorganisms in an unstained state. The special features include a phase-contrast condenser and a phase – contrast objective. Thus the special optical system makes it possible to distinguish unstained structures which differ in their refractive indices slightly.   In this objects or specimen appear dark against illuminated background.

1.1.5   Fluorescent Microscope

This microscope is used primarily for the detection of antigen –antibody reactions. The antibodies are conjugated with a fluorescent dye that becomes excited in the presence of UV light and the fluorescent portion of the dye becomes visible against a dark background. It is most commonly used to observe the specimens that are chemically tagged with a dye. A special filter is fitted to the ocular lens which permits only longer ultra violet radiations and blocks shorter ultraviolet radiations and the source of illumination being mercury lamp or hydrogen quartz lamp.

1.1.6   Electron Microscopy

Electron microscopes are scientific instruments that use a beam of highly energised electrons to examine objects on a very fine scale. Morphology of objects such as size and shape of particles, their topography, composition of elements and also the arrangement of atoms (crystallographic) in the object will be studied. The light microscopes are limited by physics of light to 500 to 1000 × magnifications and a resolution of 0.2 micrometers. Due to this limitation, electron microscopes were developed with a desire to see fine details of interior structures or organic cells viz., nucleus, mitochondria, ribosomes etc. which requires 10,000 × plus magnification that has not been possible with light microscopes. The first electron microscope prototype was built in 1933 by the German engineers Ernst Ruska and Max Knoll which was patterned exactly on the light transmission microscope except that a focused bean of electrons is used instead of light to see through the specimen.

Table 1.1 Magnification power of lenses.

Magnification		Total Magnification produces
Objective lenses	Ocular lens	
Scanning 4x	10x	40x
Low-Power 10x	10x	100x
High-Power 40x	10x	400x
Oil- Immersion 100x	10x	1000x

Working Principle

- Using the electron gun, a stream of electrons is formed and accelerated toward the specimen using a positive electrical potential. A monochromatic beam is produced using metal apertures and magnetic lenses.

- Using a magnetic lens, beam is focused onto the sample and interactions occur inside the irradiated sample. These interactions are detected and transformed into an image.

(a) *Transmission Electron Microscope (TEM) :* Using TEMs the size, shape and arrangement of the particles which make up the specimens as well as their relationship to each other and also the arrangement of atoms in the specimen and their degree of order will be studied. It works much like a slide projector. A projector shines a beam of light through the slide, as the light passes through it is affected by the structures and objects on the slide. These effects result in only certain parts of the light beam being transmitted through certain parts of slide. This transmitted beam is then projected onto the screen, forming an enlarged image of the slide.

The electron gun present at the top produces a stream of monochromatic electrons. By using condenser lenses 1 and 2, this stream is focused to a small, thin, coherent beam and is further restricted by condenser aperture knocking out high angle electrons, then the beam strikes the specimen and parts of it are transmitted. This transmitted portion is focused by the objective lens into an image. The image is passed down the column through intermediate and projector lens, enlarging the image and finally image strikes the screen, allowing the viewer to see the image.

Transmission Electron Microscope

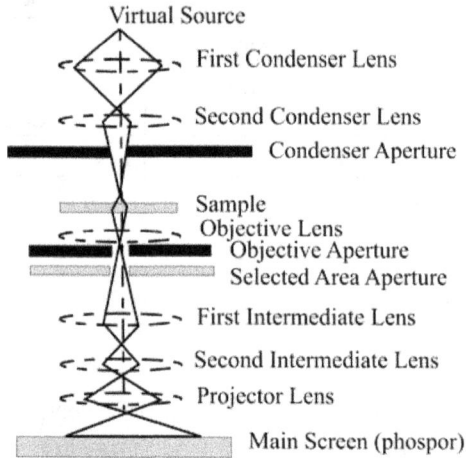

Virtual Source
First Condenser Lens
Second Condenser Lens
Condenser Aperture
Sample
Objective Lens
Objective Aperture
Selected Area Aperture
First Intermediate Lens
Second Intermediate Lens
Projector Lens
Main Screen (phospor)

(b) ***Scanning Electron Microscope (SEM) :*** Using SEMs, the surface features of an object, its texture, the size, shape and arrangement of particles making up the object that are lying on the surface of sample will be studied.

The electron gun produces a stream of monochromatic lens present at top which is condensed by the first condenser lens. It works in conjunction with the condenser aperture to eliminate the high – angle electrons from the beam. The second condenser lens forms the electrons into a thin, coherent beam. Further, objective aperture eliminates high – angle electrons from the beam. Then, a set of coils scan the beam in a grid fashion as in case of television. Finally, objective lens focuses the scanning beam onto the specimen. Interactions occur inside the sample, when the beam strikes the sample and these are detected by the various instruments.

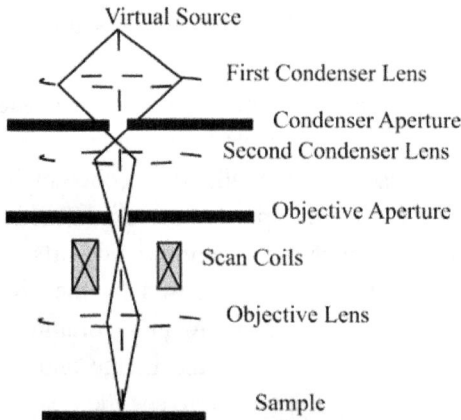

Virtual Source
First Condenser Lens
Condenser Aperture
Second Condenser Lens
Objective Aperture
Scan Coils
Objective Lens
Sample

(c) **Immuno-sorbent Electron Microscopy (ISEM)**It is used as an effective and widely applicable diagnostic tool where a combination of electron microscope and serology is involved. In this technique, virus and antiserum are reacted together. Antigens are trapped onto grid coated with specific antiserum negatively stained and the results viewed in the electron microscopy. This technique is used for different kinds of viral preparations when treated with antibodies forming decorations. This is widely used as a diagnostic tool to study those viruses which may be present in low concentrations.

Most of pathological exercises are aimed at controlling microorganisms under artificial conditions by maintaining different sterilization methods. The major physical agents used for control of microorganisms are temperature and filtration. The practical procedures by which heat is employed are conveniently divided into two categories are moist heat and dry heat.

1.2 Autoclave

The laboratory apparatus designed to use steam under regulated pressure is called an autoclave, is a double-jacketed steam chamber equipped with devices which permit the chamber to be filled with saturated steam and maintain a designated temperature and pressure for a specified period. Many culture media, discarded cultures and contaminated materials are routinely sterilized with this apparatus. The autoclave is operated at a pressure of 15 b/ cm^2 (at 121^{0}C) for 15 minutes.

Precaution

- It is essential to see that air in the chamber be completely replaced by saturated steam otherwise it results in insufficient sterilization.

1.2.1 Hot air Sterilization

Hot air oven is used for sterilizing glassware such as Petridishes, pipettes etc. It is achieved by exposing them at a temperature of 160^{0}C for one and half hour to kill vegetative mycelium and viable spores due to destructive oxidation of the cell contents.

Precautions

- The contents present inside the chamber should not be taken out by opening the door immediately after sterilization as it leads to breakage of glassware.
- Plasticware and other heat labile materials should not be sterilized using hot air oven.

1.2.2 Filtration

A variety of filters have been available to plant pathologists which aid in removing microorganisms from liquids or gases. These filters are made of different materials such as an asbestos pad in the seitz filter, sintered glass disks in sintered glass filter, or porcelain in chamberland filter. These are used for sterilizing the sensitive materials like vitamin solutions, antibiotics etc.

1.2.3 Laminar Air Flow System

With the development of high efficiency particulate air (HEPA) filters, it has been possible to deliver clean air to an enclosure such as a cubicle or room. Laminar air flow is extensively used to produce dust and bacteria free air. It is used for aseptic transfer of microbial cultures under sterilized conditions.

1.3   BOD Incubator

Microorganisms can grow over a wide range of temperatures, from very low temperatures which is characteristic of psychrophiles to the very high temperatures, characteristic of thermophiles and optimum temperatures, characteristic of mesophiles. Optimum growth of microbial growth is obtained by providing ambient temperature conditions under controlled conditions using an incubator. Generally a temperature of 25°C-28°C is maintained for optimum growth of fungi where as slightly high temperature conditions i.e., 30-32°C are used for growth of bacteria.

1.4   Spectrophotometer

It is a very useful instrument for obtaining a relative estimate of cell number or cell mass based on the turbidity of culture. It works on the principle of Beer-Lamberts law. This instrument measures the amount of light transmitted or absorbed. It transmits a beam of light at a single wave length through a liquid culture. The cells suspended in the culture interrupt the passage of light, and the amount of energy transmitted through the suspension is measured on a photoelectric cell and converted into electrical energy. The electrical energy is then recorded on a galvanometer as 0 to 100% transmittance.

1.5   Centrifuge

Centrifuge is used for separation of microbial cells such as fungi, bacteria, viruses from the infected material based on centrifugal force. Different types of centrifuges available based on rpm. Ultracentrifuge may have speed as high as 15,000 rpm or even more.

1.6 PCR Thermal Cycler

It is deployed to carryout polymerase chain reaction, a powerful technique that has widespread applications in molecular biology. This technique is used to amplify a specific nucleic acid fragment that lies between two regions of known nucleotide sequence, and often from an extremely small amount of target nucleic acid in biologically complex samples. PCR-based assays are more advanced for the detection of important viral, bacterial and fungal pathogens of many cultivated crops. PCR methods are also used to detect and characterize plant viruses, viroids ad phytoplasmas. Because of the sensitivity of PCR, viral infection which cannot be detected by serological methods can readily detected with PCR methods.

1.7 ELISA Plate Reader

It is used to carryout ELISA (Enzyme linked Immunosorbent Assay) test, a commonly used method to detect and identify a pathogen particularly viruses. ELISAs are highly sensitive, allow for large sample numbers to be processed at the same time, require small quantities of antiserum. ELISAs come in several formats. The most common are double antibody sandwich (DAS) ELISA and DAC (double antigen coated) ELISA.

1.8 Chromatography

It is one of most widely used analytical methods for seperation of proteins and mycotoxin analysis. Especially for mycotoxin analysis, various analytical methods have been developed such as Thin Layer Chromatography (TLC), High performance Thin Layer Chromatography (HPTLC), Paper chromatography etc.

1.9 Photomicrography

The recent trend in studying microscopic features of fungi and bacteria is through software aided computer programming. For this purpose, there will be an attachment of camera to holder of eye piece of a binocular research microscope through a cable wire and further the images are captured by viewing on the computer monitor. A specially designed software programme for the purpose is needed that will be supplied by different firms. It is relatively easy to focus and capture the images at will. This is a recent addition in the image processing rather than the usual way of photomicrographs taken with the help of trinocular microscope. There are several firms like Olympus Magnus, Zeiss Axiomat which are supplying the device. The contrast, brightness, sharpness of image can be adjusted with the aid of software features available. It is one of the very useful aid for plant pathologists to study about different structures of fungi, bacteria, actinomycetes and nematodes.

1.10 Orbital Shaker

It is used for production of suspension cultures in tissue culture studies. Suspension culture is initiated by first transferring an inoculum of friable callus to a liquid medium and then placing it on a mechanical orbital platform shaker at 100-150 rpm. Agitation is required for suspension cultures so as to ensure uniform distribution of cells of various sizes and shapes and provides gas exchange for the cells to sustain cell respiration in the medium.

1.11 Colony Counter

It is used to count number of bacterial colonies produced on a solid medium like nutrient agar medium. A digital display is provided to show the total number of colonies counted automatically.

1.12 Refrigerator / Deep Freezer

It is essential to preserve fungal and bacterial cultures at low temperatures. It is also used to store stock solutions of vitamins, hormones needed for preparation of various tissue culture media.

1.13 Lyophilizer

It is used for preserving microbial cultures for 15 to 20 years without any physiological change and loss of genetic stability. It is mostly used in culture collection centres and industries for preservation of microbes.

Chapter 2

Isolation, Purification and Preservation of Microbial Cultures

Plant pathogens are the major culprits responsible for a menace in larger parts of world. They induce different diseases on various crop plants. For the purpose of identification and differentiation from closely related species, they have to be isolated in pure culture and their morphological characters, physiological and biochemical activities are to be studied. It is often noticed that, where one pathogen got mixed or contaminated by more than one organism. Hence, purification of microbial cultures assumes utmost importance.

2.1 Types of Media

Microorganisms are diverse in their nutritional requirements. There are different types of media available for isolation under artificial conditions. There is no universal medium to support growth of all microorganisms. Hence, a clear understanding of the nutritional requirements of a specific microbe or a group of microbes is essential to constitute a suitable medium.

Based on **chemical composition**, the media can be broadly classified into
- (a) Natural medium (or) Non-Synthetic medium
- (b) Semi – Synthetic medium
- (c) Synthetic medium.

2.1.1 Natural (or) Non-Synthetic Medium

In this, the percent of elemental composition of all chemicals is not completely known. Most of the fungi grow better on media containing natural substances than the synthetic medium. But the disadvantage being with this is no uniform or consistent growth of them because of the difference in composition of nutrients.

Ex : Apple juice agar, carrot juice agar, malt extract agar etc.

2.1.2 Semi – Synthetic Medium

Semi synthetic media are those whose percent elemental composition of one or two components is partially known

Ex : Yeast extract dextrose agar

13

2.1.3 Synthetic Medium

The chemical composition of all elements is completely known. Based on the purpose and requirements the preparation of media depends.

Ex : Peptone rose bengal agar medium, Czapeck's dox agar medium.

Based on **consistency,** the media can be classified into liquid, solid and semi-solid medium.

2.1.4 Liquid Medium

It is used to study biochemical characteristics of pathogens. But using this medium, it is not possible to study the morphological characters of the organism concerned.

Ex : Potato dextrose broth, Nutrient broth.

2.1.5 Solid Medium

This is prepared by adding solidifying agent such as gelatin or agar to a corresponding liquid. If agar is added to potato dextrose broth or nutrient broth, it becomes potato dextrose agar and nutrient agar, a solid medium respectively. Agar-agar is the most convenient solidifying agent because it melts around $98^{\circ}C$ and does not solidify until cooled to about $42^{\circ}C$. Therefore, an addition of 15-20 grams of agar in one litre of a liquid will produce a gel. A semisolid medium with agar at 0.5% concentration or less will have a soft custard like consistency and is useful for the cultivation of microaerophilic bacteria or for studying bacterial motility.

Irrespective to the category of medium, it is possible to isolate specific plant pathogens using selective medium

Ex : Trichoderma selective medium – For isolation of *Trichoderma* spp.

King's B Medium – For isolation of *Pseudomonas fluorescens*

2.2   Isolation of Fungal Pathogens

The fungi, infect different plant parts and produce visible manifestations of disease soon after the completion of incubation period in the host plant. The fungal pathogens can be isolated from infected plant tissues such as leaves, stem, fruits, root etc to study the morphological characters which form the basis for identification of various fungal genera.

2.2.1 Isolation from leaves

The infected leaves are thoroughly washed in sterile distilled water with the help of fine blade or sharp razor, the infected tissues along with adjacent small unaffected tissues are cut into small pieces of 2- mm^2 and are transferred to sterile Petridish containing 0.1% $Hgcl_2$ with the help of sterilized forceps. Besides mercuric chloride, sodium hypochlorite (1%), hydrogen peroxide can also be used as surface sterilants. The infected cut pieces are surface sterilized in 0.1% $Hgcl_2$ for 30-60 sec. Time of exposure will vary with plant part to be sterilized. After that, it is subjected to wash in sterile distilled water for two to three times. Then, the leaf bits are transferred aseptically to pre-sterilized Petridishes containing potato dextrose agar supplemented with streptomycin sulphate to suppress the bacterial contamination. Keep the plates in an incubator at optimum temperature of 25-27°C.

2.2.2 Isolation from Other Parts

The fungal pathogens from stems, roots or fruits in which they may present in deep seated tissue have to be isolated by culturing pieces from internal tissues. The infected tissues are thoroughly washed in sterile water and then swabbed with cotton wool dipped into 90% ethanol followed by exposure to an alcohol flame for few seconds. The outer layer of tissues are quickly removed by a sterilized scalpel and sterilized by dipping into 90% ethanol then flaming for few seconds. The tissues are thus sterilized, transferred to nutrient agar poured in Petridishes and incubated. The fungal mycelium growing from infected tissues is transferred to agar slants kept in tubes.

2.3　Isolation of Bacterial Pathogens

Bacterial pathogens infecting leaves and other plant parts are generally isolated by preparing a bacterial suspension. Infected leaves showing symptoms of bacterial infection are cut into small bits and surface sterilize with 0.1% $Hgcl_2$ or 70% ethanol for one minute and wash separately in distilled water. These surface sterilized infected bits are taken into a test tube and add few drops of sterile distilled water and crush with help of sterilized glass rod. Then the suspension is streaked on nutrient agar medium poured plates and incubate at 30°C for 24-28 hrs. The discrete individual bacterial colonies are transferred to agar slants.

2.3.1 Pure Culture Techniques

There are 3 methods available for purification of bacterial cultures

- (a)　Pour plate method
- (b)　Streak plate method
- (c)　Spread plate method

Among these techniques, streak plate method is most commonly used in microbiological laboratories for purification of bacterial cultures.

2.4   Purification of Fungal Pathogens

When once an organism isolated, it has to be purified for its identification and characterization. For purification of fungi, two methods are available such as

- (a) Single hyphal tip method
- (b) Single spore isolation

2.4.1 Single Hyphal Tip Method

A small disc of agar medium containing fungal growth is transferred to the centre of potato dextrose agar medium (PDA) using a sterilized inoculation needle and incubate the plates at optimum temperature for few days. As the fungus grows, the advancing edge of the fungal growth will have well separated hyphal tip, which are marked off by a glass marking pencil observing the bottom of petridish. Those bits are cut with the help of cork borer and transferred to agar slants individually and allow to grow keeping in an incubator and the hyphal tips would give rise to pure cultures.

2.4.2 Single spore Isolation

Take small amount of fungal mycelial mat of sporulating fungi (*Fusarium, Alternaria*) and add to a test tube containing 9 ml of sterile distilled water. From that, prepare serial dilutions upto 10^{-5}. From 10^{-5} dilution, take 1ml of suspension and add to plates containing water agar (2%) medium. Locate the spores under low power objective by reversing the petriplates Encircle the spotted spores and keep it for incubation at $27^{0}C$ and observe for germination of spores at periodical intervals. After two days, cut the disc containing germinating spore with the help of cork borer and transfer to PDA slants and incubate. From the single spore, pure culture of fungus would grow.

2.5   Maintenance and Preservation of Microbial Cultures

The transfer of organisms to fresh media at regular intervals is time consuming and laborious. Many bacteria and fungi become adapted to saprophytic growth, loses genetic stability and viability and often fail to sporulate. Morphological changes can create confusion in identification. Selection of a preservation method is based on nature of pathogen. If the pathogen is not well understood, preservation by more than one method should be done. There are several methods available for preserving microbial cultures, are briefly described here.

2.5.1 Agar Slant Cultures

The general practice of maintaining a culture is to inoculate the organism in agar slants / stabs and to incubate the culture for a short period thus provide sufficient time for the organisms to grow. The minimum incubation and subsequent refrigeration prevent accumulation of metabolic products in the media and thus allow them to survive for longer periods. When a culture is preserved by refrigeration, it is necessary to transfer the culture to a fresh agar slant at intervals of few weeks or months depending on the nature of a culture. Due to frequent subculturing, cultures may become genetically unstable and show changes in some characteristics. However, the method is inexpensive and suitable for routine pathological laboratories.

2.5.2 Agar Stab Method

By this technique, especially microaerophilic bacteria can be preserved as it requires low concentration of oxygen for its growth and atmospheric oxygen concentration may be injurious to them. In this method, a tube of sterile agar medium is solidified in upright position. Following which the medium is inoculated by thrusting culture with the inoculation needle into the centre of the agar stab. During incubation, a core of growth develops. Strict anaerobes are cultured by this technique and preserved in a strongly reducing medium.

Frequenting the transfer of fungal culture is the simplest method for maintenance of living fungi and is achieved by frequent or serial transfer from stated media to fresh suitable solid or liquid media and storage in most suitable conditions for the individual isolate. But the disadvantages associated are

(i) Loss of pathogenicity or other morphological or physiological properties.

(ii) Loss of genetic stability and danger of variation.

(iii) Problem of contamination either by air borne spores or mite carried infections.

(iv) Requires constant supervision by specialists.

The time interval between transfers varies from fungus to fungus, for some it may be required once every two to four weeks while others may survive for 6-12 months without any transfer. Preserving cultures under low temperature can slow the rate of metabolism and thus increase the intervals between subcultures.

2.5.3 Storage under Mineral Oil

This is the most extensively used method of preserving fungi. Agar slants containing healthy full grown, mature cultures are overlaying with mineral oil, at least 1 cm above the tip of the slant. The mineral oil used must be sterile and this can be achieved by alternate heating and cooling it twice in an autoclave at 121°C for 15 minutes. This method minimizes dehydration, probability of contamination and also the frequency of subculturing. Retrieval of fungal culture is attained by draining away as much oil as possible and a small amount of fungal mycelium is removed with inoculation needle and inoculating onto a suitable nutrient medium. After retrieval, growth rate often reduced due to adhering oil and more than one subculture may be necessary. The fungal mycelium can normally recover when it is re-isolated from the agar plated onto fresh medium. A wide range of fungi survive by this method. The drawbacks of oil storage are (i) there will be always reduced growth on retrieval (ii) chances of contamination by air-borne spores are more.

2.5.4 Storage in Silicagel

This method has proved to be very successful in preserving sporulating fungi for about 7 years at CMI. The cultures remain morphologically stable after retrieval.

Universal glass bottles are quarter filled with non-indicating silica gel and sterilized in an oven at 180°C for 3 hrs. The bottles are placed in a tray containing water and subjected to deep freeze. Spore suspensions are made in cooled 5% skimmed milk and about three quarters of silica gel is wetted with spore suspension and incubate at 25°C. After 10-14 days of incubation, when the crystals readily separate, the bottles are stored at 4 °C in an air tight container containing self indicating gel to absorb moisture. A wide range of sporulating fungi are preserved by this method.

Limitation

(i)   Thin wall spores, spores with appendages does not survive by this technique.

2.5.5 Lyophilization (or) Freeze Drying

Most of the microbial cultures die when cultures become dry. However, freeze drying can satisfactorily preserve many kinds of fungi and bacteria that would be killed by ordinary drying. In this process, a dense spore suspension is prepared in 10% skimmed milk and 5% Inositol which are sterilized by autoclave. The spore suspension is placed in small vials and frozen at −60°C to −78°C. The frozen cultures are dried under vaccum which results in dehydration of cultures with minimum damage to delicate cell structures. The vials are then sealed off under vaccum and stored in a refrigerator. The ice present in frozen suspension sublimes under vaccum i.e. evaporates without first going through a liquid water phase. Cryoprotective reagents such as glycerol or dimethyl sulfoxide (DMSO) in the medium, prevents cell damage due to i.e., ice crystal formation. Many species of fungi and bacteria are preserved by this method which remain viable and genetically stable for more than 20 years. This method requires minimum storage space and it is easy to transport cultures. Generally, lyophilization is used in culture collection centres and industries.

Limitation

(i) It requires costly equipment.

2.5.6 Liquid Nitrogen Storage (or) Deep Freezing

This method employs ultra low temperatures to cease the metabolism of microorganisms. It can be widely applied to sporulating and non-sporulating cultures of fungi which are kept stable throughout their storage period. The method includes pre-cooling of 0.5 ml aliquots of fungal suspension in 10% glycerol sealed in 1 ml borosilicate glass ampoules to 7°C and freezing at a rate of 1°C /min. at –35°C. Then the ampoules or vials are then stored in a liquid nitrogen refrigerator either by immersion in liquid nitrogen (-196°C) or by storage in the gas phase above the liquid nitrogen. The liquid nitrogen method has been successful with many fungal and bacterial species, which remain viable for 10 to 30 years without under going genetic change in their characteristics. It is necessary to replenish the liquid nitrogen at regular intervals to replace the liquid nitrogen lost due to evaporation.

2.5.7 Water Storage

This method of storage was originally used to store fungi pathogenic to man. Plant pathogens including Phycomycetes, Ascomycetes, Basidio-mycetes and Fungi imperfecti could be stored which remain viable and pathogenic for 7 years. Agar blocks of 5 mm (mycelial discs) are cut from the growing edge of a fungal colony and the blocks are placed in sterile distilled water in Mc Cartney bottles and the lids are to be tightly screwed down and store at room temperature.

2.5.8 Soil Storage

This method of storage has proved very successful with soil-borne fungi like *Fusarium* spices, molds viz., *Aspergillus, Penicillium, Rhizopus, Alternaria* species for about 5 years without any morphological or physiological change. The method involves inoculation of double autoclaved soil with 1 ml of spore suspension in sterile distilled water and incubating at room temperature for 5-10 days depending on growth rate of fungus being stored. This initial growth period allows the fungus to use the available moisture and then they are stored in a refrigerator at 4-7 °C.

Chapter 3

Identification of Some Commonly Occurring Plant Pathogenic Fungi

In Plant Pathology, accurate identification of different fungal genera occupies paramount importance. It has always been remained as a tough and daunting task which can never be pursued without perseverance. In this chapter, a list of fungi which are often encountered are described here along with diagrams.

3.1 Downy Mildews (Peronosporales)

Downy mildew fungi are obligate parasites, cosmopolitan in nature. Appearance of white powdery growth on any substrate, a growth on plant surface, often on a leaf, which is superficial in nature. Mycelium is intercellular in nature produce haustoria. Asexual reproduction is by means of 'Zoospores' except in *Perenospora* in which the spore is a conidium. Sporangiophores or conidiophores are branched can be differentiated from the mycelium and determinate, emerge mostly through stomata. Sexual reproduction is by means of oospores formed due to the fusion of oosphere being fertilized by nuclei in an antheridium.

Colonial Morphology & Microscopic features	Diagram
(i) *Perenospora* Conidiophore is usually dichotomously branched, tips pointed conidia germinate by germ tube. Unlike other genera in the family, the asexual spore does not form zoospores. The rate and distance of spread within a crop are much greater than in genera where zoospores are formed. **Ex :** *P. anemones;* *P .antirrhini*	

Contd...

Colonial Morphology & Microscopic features	Diagram
(ii) Plasmopara Sporangiophores are formed in clusters usually on lower side of the infection spot. Branches of sporangiophores arise typically at right angles and irregularly spaced, which ends with final branches known as sterigmata bearing sporangium. Germination usually by means of zoospores. Ex : P. Viticola	
(iii) Basidiophora The fungus characteristically produces sporangiophores which are unbranched, club shaped (basidium like) with swollen heads on which papillate sporangia are present. **Ex :** *Basidiophora butleri, B. kellermanii, B.entospora*	
(iv) Pseudoperenospora Unlike *Perenospora*, in this vegetative reproductive unit is a sporangium forming zoospores; sporangium does not develop into germ tube. **Ex :** *Pseudoperenospora cubensis*	
(v) Sclerospora Sporophore shows dichotomous branching in the upper part, sporangia forming zoospores, non-papillate, oogonial wall thick, oospore plerotic. *Sclerospora* could be divided into two sections based on germination such as	

Contd...

Colonial Morphology & Microscopic features	Diagram
(a) *Eusclerospora* (b) *Perenosclerospora.* In *Eusclerospora,* those with a sporangium which germinate by zoospores. In *Perenosclerospora,* those with a conidium germinating by a germ tube.	
(vi) *Bremia* Sporophores branch with acute angles, dichotomous branching, with enlarged tips, like a disc and bear 3 or 4 projections, sterigmata are characteristic spores with an apical papilla and germinate directly i.e. not by zoospores. **Ex** : *Bremia graminicola B. lactucae* The genus *Bremiella* differs *Bremia,* in that the tips of the sporophore branches are blunt. **Ex** : *Bremiella baudysii,* B. *megasperma*	

3.2 Pin Molds (Mucorales)

Members of Mucorales are commonly known as pin molds as their sporangia appear as black dots in the cobweb-like hyphae. Also called as 'sugar fungi' because of their involvement in decaying vegetable matter, utilize simple carbohydrates and convert it into complex carbohydrates such as cellulose, hemicellulose etc. Reproduce asexually by non-motile aplanospores usually produced in sporangia and sexually by gametangial copulation, result in the formation of zygospores.

Colonial Morphology & Microscopic features	Diagram
(i) Rhizopus Colonies are greyish brown, cover the petriplates in 4 days on PDA at 25°C Good growth and sporulation at 15-40°C, no growth at high temperature (at 45°C). Rhizoids brownish, sporangiophores arise on stolons, sporangia greyish black, powdery in appearance sporangiospores angular, subglobose to ellipsoidal, with ridges on the surface. **Ex : *Rhizopus oryzae***	
(ii) Mucor Colonies light greyish, covering the petriplates in four days, composed of tall and small sporangiophores. The growth and sporulation are good at 20-25°C and poor at 15°C and 37°C. Tall sporangiophores sympodially branched, short sporangiophores profusely branched with circulate branches, sporangia with incrusted walls, coumellae obovoid to ellipsoid in larger sporangia and globose in smaller sporangia. **Ex : *Mucor circinelloides***	
(iii) Syncephalastrum Colonies are white at first later turn grey, about 6 mm in height. Sporangiophores unbranched at first, later richly branched. Fruiting head oval or globose, with numerous small warts to which the mersoporangia are attached, merosporangia 5-10 spored. **Ex : *Syncephalastrum racemosum***	

Contd...

Colonial Morphology & Microscopic features	Diagram
(iv) *Cunninghamella* Colonies grow rapidly at 25^0C, powdery in appearance, white at first, later become pale smoke grey, mycelium hyaline. Conidiophores erect, hyaline. Vesicles globose to subglobose. conidia globose, echinulate, heterothallic. **Ex :** *Cunninghamella echinulata*	

3.3 Powdery Mildews (Erysiphales)

Usually a superficial, whitish powdery growth produced on living plants by fungi, obligate parasites, Ascomata cleistothecial, anamorphs hyphomycetous, prominent. **Ex :** *Oidium, oidiopsis, Ovulariopsis.*

Colonial Morphology & Microscopic features	Diagram
(i) Erysiphe It is characterized by perithecia with many asci. Mycelium superficial forming haustoria in the epidermal cells of the host, conidiophores produce conidia in chains, conidia, single celled, perithecia non-ostiolate, myceloid appendages, asci many, ascospores single celled. **Ex :** *Erysiphe polygoni*	

Contd...

Colonial Morphology & Microscopic features	Diagram
(ii) *Phyllactinia* Characterized by its perithecia with many 2-spored asci. The anamorph also differs from that of other genera markedly and belongs to the genus *ovulariopsis*. Mycelium external and internal, forming haustoria in parenchymatous cells of host. Conidiophores producing conidia singly, conidia hyaline, single celled. Asci many, Stalked; ascospores 2 to 3, appendages with bulbous base. **Ex :** *Phyllactinia guttata*	
(iii) *Podosphaera* Characterized by its perithecia with one ascus containing 6-8 spores, appendages dichotomously branched at the apex. The genus is distinctly separated from *Microsphaera* with the same kind of appendages, but with many asci. *Podosphaera* is also clearly delimited from *Sphaerotheca* which is also with one ascus, but with unbranched or irregularly branched appendages. Mycelium superficial, haustoria formed in epidermal cells of host. Conidiophores producing conidia in chains, conidia hyaline, single celled, perithecia non-ostiolate, appendages dichotomously branched in the apex, asci one, ascospores 6 to 8. **Ex :** *Podosphaera myrtilliana*	

Contd...

Colonial Morphology & Microscopic features	Diagram
(iv) *Microsphaera* Characterized by its perithecia with many asci and dichotomously branched appendages. The genus is separated from *Podosphaera* having same kind of appendages but with only one ascus. **Ex** : *Microsphaera divaricata.* Mycelium superficial, formation of haustoria in the epidermal cells of host, conidiophores producing conidia singly, appendages are dichotomously branched in the apex, asci many, stalked, ascospores 3-8	
(v) *Sphaerotheca* Characterized by its perithecia with one ascus containing 6-8 spores with myceloid appendages the genus is separated from *Erysiphe,* contain myceloid appendages but with many asci, as well as from *Podosphaera* with one ascus but dichotomously branched appendages. **Ex** : *Sphaerotheca pannosa*	
(vi) *Uncinula* Characterized by its perithecia with many asci and appendages are coiled apically. Mycelium superficial, haustoria formed in the epidermal cells of host, conidiophores producing conidia singly, conidia hyaline, single celled, appendages coiled apically asci many, stalked, ascospores 3-8. **Ex** : *Uncinula bivonae, U.necator*	

Contd...

Colonial Morphology & Microscopic features	Diagram
(vii) *Leveillula* Appendages myceloid, simple or irregularly branched. Anamorph – oidiopsis. Mycelium endophytic and ectophytic, conidiophore simple or branched, conidia formed singly. Asci numerous ascospores large. **Ex :** *Leveillula taurica*	

3.4 Rust Fungi (Uredinales)

Rust-fungi belong to the order Uredinales, sub-division Basidiomycotina, distributed all over the world parasitizing a wide range of host plants occuring both in temperate and tropical region. Majorly three types of rust cycles exist such as A. macrocyclic rust, in which at least three or four spore stages occur during its life cycle namely, Pycnidium (O), Aecium (I), Uredium (II), Telial (IV) and Basidial (V) states. This phenomenon is known as polymorphism or Pleomorphic. B. Demicyclic rust, where uredial stage is lacking C. Microcyclic, in which atleast one of the important spore stage must occur either uredial or telial.

Colonial Morphology & Microscopic features	Diagram
(i) *Puccinia* Aecia subepidermal in origin, uredinioid with peridium and echinulate, erumpent, spores borne singly on pedicels. Uredinia without peridium but may have paraphyses. Uredospores borne singly on peridium. Echinulate. Teleutospores typically 2-celled by horizontal septum, borne singly as pedicels. Ex : *Puccinia graminis*	

Contd...

Colonial Morphology & Microscopic features	Diagram
(ii) *Hemileia* The genus is characterized by uredia, typically reni-form shaped, suprostomatal but in some species erumpent through the ruptured epidermis. Spores borne singly on short pedicels, strongly asymmetrical with 1 flat or concave side and a dorsal, convex, echinulate. Spermogonia and aecial stages are unknown. Telia as the uredinia of the species, single celled, borne singly on short pedicels. **Ex** *: Hemileia vastatrix*	
(iii) *Uromyces* Aecia erumpent, aecidioid with peridium and subepidermal in origin. Uredinia subepidermal, erumpent, spores borne singly on pedicels usually echinulate. Teliospores borne singly on pedicels which may or may not be grouped. **Ex** : *Uromyces appendiculatus*	
(iv) *Ravenelia* Characterized by presence of telial heads uredinoid with spores, pedicellate. Uredinia mostly subepidermal, often paraphysate, spores borne singly on pedicels, mostly echinulate. Telia subepidermal. Mostly 1-celled, discoid heads, surface smooth or sculptured. **Ex** : *Ravenelia glandullosa*	

Contd...

Colonial Morphology & Microscopic features	Diagram
(vi) Melampsora Caemaoid type of aecia. Spermogonia subcuticular, uredinia bright yellow or orange when fresh with abundant paraphyses, Teleutosori, forming crusts consisting of a single layer of spores, single celled, sessile. **Ex :** *Melampsora euphorbiae*	

3.5. Smut Fungi (Ustilaginales)

Smut fungi belong to Ustilaginales of Basidiomycotina. The classification of smut fungi is chiefly based on teliospore germination i.e. terminal or intercalary.

Colonial Morphology & Microscopic features	Diagram
(i) Sphacelotheca It causes systemic infection. The sori in infected ovaries covered by peridium which is composed of pericarp tissues and fungal cells and liberates black powdery mass and the central columella. The columella is composed of mainly fungal cells. Spores are sub-globose to ovoid in shape, light to medium reddish brown, formed from the base of sori in chains. The spores are held together by **disjunctor cells** when at young stage. Spores on germination results in the formation of four-celled basidia with laterally and terminally produced basidiospores. **Ex :** *Sphacelotheca sorghi*	 Sphacelotheca Teliospore

Contd...

Colonial Morphology & Microscopic features	Diagram
(ii) *Ustilago* Sori contain blackish-brown powdery mass in affected spikelets. Sori covered by a membrane of host origin and hence called **covered smut** initially later exposed to various degrees depending on the extent of destruction of the spikelet parts. Spores lighter coloured on one side, smooth, globose, subglobose to ovoid. Dikaryons will be formed compatible basidial cells or basidiospores, thus infect seedlings. Spore germination results in a four celled basidium on which ovoid basidiospores are produced on artificial media. **Ex :** *Ustilago hordei*	 Teliospore
(iii) *Urocystis* Sori are formed in different plant parts such as in deformed leaves, sheaths, culms and spikes, Sori appear as long, sometimes confluent streaks. Initially sori are covered by epidermis upon rupture longitudinally leads to liberation of reddish-black dusty mass of spore balls. Spore balls are subglobose, ovoid to irregular and surrounded by an incomplete layer of sterile cells. Spores are variable in size and shape. Spores upon germination results in the production of aseptate basidia of variable length, bearing 4-6 apical and elongated basidiospores. **Ex :** *Urocystis pompholygodes*	 Fertile cells Sterile cells

Contd...

Colonial Morphology & Microscopic features	Diagram
(iv) *Tolyposporium* The genus is characterized by formation of black, agglutinated to granular spore mass composed of ***spore balls*** in a ***tumor like sori***. It is mostly localized in flowers and parts of flowers. Spores variable in size and shape such as subglobose, elongated or irregular, spore balls composed of 3-40 or more spores. Spores undergo a period of dormancy. Spore germination results in the formation of four celled basidia with basidiospores borne at the septa after a period of resting usually in the presence of water. **Ex** : *Tolyposporium junci*	
(v) *Melanopsichium* The fungus usually affects inflorescence, however perceptible irregularly lobed galls also found on stems and leaves. The galls are composed of hypertrophied host tissue containing numerous, spherical to irregular chambers filled with spores embedded in a hyaline, gelatinous matrix. In humid conditions, this gelatinous substance swells considerably and pushes the spore mass out to the sorus surface. Spores globose, ovoid to some what irregularly elongated. Germination of spores results in formation of one or two basidia on which basidiospores are borne.	

Contd...

Colonial Morphology & Microscopic features	Diagram
(vi) *Tilletia* The genus is characterized by formation of **H-shaped, filiform or elongated basidiospores** produced apically upon spore germination which are 8-16 in number. The infected ovaries filled with agglutinated spore mass. Basidiospores fuse to form dikaryotic hyphae which infect seedlings. **Ex :** *Tilletia caries*	H-shaped pieces
(vii) *Neovossia* The fungus usually occur at the time of inflorescence, inconspicuous sori, on rupture release black,. semi-agglutinated to powdery spore mass. Spores are ovoid to irregularly elongated. Germination of spores results in holobasidia which bears a large number of **fusiform or slightly bent basidiospores** terminally. **Ex :** *Neovossia indica*	

3.6 Mushroom fungi (Agaricales)

The science of Mycology started with the study of mushrooms. Mushroom fungi belong to Agaricales of Basidiomycotina. The edible species are called **'mushrooms'** where as poisonous ones are called **'Toad stools'**. Because of its high nutritive valvue, lot of economic importance is associated with these fungi.

Colonial Morphology & Microscopic features	Diagram
(i) *Boletus* Pileus round, reddish brown in colour resembles **doughness** of a **bread**. Stem with characteristic reticulate used for culinary purpose. The genus usually found growing on soil in woods. **Ex :** *Boletus edulis*	
(ii) *Amanita* Poisonous and non-edible mushroom. It is used in killing mosquitoes and hence called "**Fly mushroom**" with characteristic scales on the surface of a pileus, presence of volva with narrow stalk which can be easily separable. It is used in preparation of intoxicating beverages. Two alkaloids namely muscarine, muscaridine are produced. **Ex :** *Amanita muscaria Amanita Phalloides*	
(iii) *Pholiota* These are **wood decaying fungi** commonly found an growing birches. Fruits bodies grow in tufts. Umbrella shaped, scales are present over the surface of pileus and below the annulus on the stipe. *P. squarrosa* is the most common species.	
(iv) *Psilocybe* Commonly called as '**sacred fungi**' pileus is conical in shape, pale-buff when dry. Stipe is long, slender and warty. The genus is associated with hallucinogenic properties due to two alkaloids namely **psilocin** and **psilocybin.** *P. semilanceata* – commonly called as **Liberty cap.**	

Contd...

Colonial Morphology & Microscopic features	Diagram
(v) *Agaricus* Basidiocarp is umbrella shaped consists of stipe, negatively geotropic and fleshy, convex pileus. Fruit bodies spread radially called as **Fairy ring**. Gills are hanging vertically from the pileus. Species of *Agaricus* grow saprophytically on decaying leaves, manure heaps and humus rich soil. **Ex :** *A. campestris*	
(vi) *Pleurotus* Popularly called as '**Oyster mushroom**' with characteristic slippery, fan shaped fruiting bodies. Stipe laterally attached, not usually found. **Ex :** *Ostreatus*	

3.7. Hyphomycetous Fungi

The class Hyphomycetes includes conidial states whose perfect state is unknown or lacking. Conidial states are grouped into form -genera for convenience in identification and nomenclature. It is part of an additional, special purpose cross-classification in which the conidial states of Ascomycotina, Basidiomycotina and sometimes certain Zygomycotina are grouped together. The class includes those fungi, **forming spores directly on the hyphae or on sporophores, but never produced in pycnidia or acervuli.**

Colonial Morphology & Microscopic features	Diagram
(i) Pyricularia Conidia **bi-celled, pyriform** borne singly at the tip of conidiophore and its successive growing points. Conidiophores are simple, erect and hyaline. **Ex :** *Pyricularia oryzae*	

Contd...

Colonial Morphology & Microscopic features	Diagram
(ii) *Alternaria* The conidia of *Alternaria* are highly characteristic, yellowish brown, **beaked** and **possess transverse** and **longitudinal septa**. Spores with this type of septation are described as **muriform**. Perfect state of Alternaria is *Pleospora*. **Ex :** *Alternaria helianthi*	 Germ Tubes
(iii) *Botrytis* **(Grey mold)** Conidia single celled, oval, apiculate at base. Spores are easily blown away by wind. As colonies grows old, sclerotia are formed after few days. Conidiophores are simple, erect. variable in length. Conidia look like **grape bunches** on conidiophores. **Ex :** *B. cinerea* – best known species (perfect state : *Sclerotinia*)	
(iv) *Monilia* Conidia produced in **basifugal fashion, oval or lemon shaped** which are often linked by is ithusmus. Conidiophores are variously branched, arise terminally or laterally from the hyphae. **Ex :** *M. fructigena* – perfect state *Monilinia*	

Contd...

Colonial Morphology & Microscopic features	Diagram
(v) Nigrospora Conidia single celled, **black** and **globular slightly bulged apically.** Conidiophores are small, flask shaped forming a single conidium at the tip **Ex :** *N .oryzae – Khuskia* (Perfect state).	
(vi) Drechslera : Conidia develop apically from a pore at the tip of conidiophore. Conidia are **cylindrical** or **club shaped** and transversely septate. **Ex :** *Drechslera sorokiniana : Cochliobolus sativus* (Perfect state)	
(vii) Curvularia Spores are usually curved, the **third cell from the base of the spore is larger than the rest** and the end cells are paler. Conidiophore forms a sympodula bearing a cluster of conidia. The perfect state of Curvularia are species of Cochliobolus **Ex:** *C.lunata.*	
(viii) Cercospora The conidia are narrow and tapering, and contain numerous transverse septa. Most species of *Cercospora* are weak parasites responsible for causing leaf spot disease in several crop plants. Conidiophores emerge through stromata or ruptured epidermis, arise in tufts from the stroma. Conidiophores with characteristic **geniculate bends** on which conidia are borne acrogenously. The perfect state of *Cercospora* are species of *Mycosphaerella* **Ex :** *Cercospora musae*	 — Conidia — Conidiophore — Host tissue

Contd...

Colonial Morphology & Microscopic features	Diagram
(ix) Fusarium The genus is an important plant pathogen causing wilt and root rots especially in tropics. Conidia are produced in fructifications called **sporodochia**. Two types of conidia are produced (a) Micro conidia – single celled, oval or comma shaped. (b) Macro conidia – spindle shaped with pointed ends, 3 to many septate with a distinct pedicellate base. Chlamydospores are produced terminally or intermittent, perfect stages are *Gibberella, Nectria* and *Calonectria.* **Ex :** *Fusarium oxysporum*	

3.8 Coelomycetous Fungi

Coelomycetous fungi belong to Fungi Imperfecti (Deuteromycotina) in which conidia are formed within a cavity lined by either fungal tissue, host tissue or a combination of both. The concept is based on **conidia production within a specified fruiting body either acervuli or pycnidia.**

Colonial Morphology & Microscopic features	Diagram
(i) Ascochyta Pycnidia contain hyaline, two celled, conidia arise by septation from the conidiogenous cell. Conidia are 2-3 celled, hyaline. Perfect state is *Mycosphaerella.* Ex : *Ascochyta rabei*	

Contd...

Colonial Morphology & Microscopic features	Diagram
(ii) Colletotrichum Conidial fructification is an **acervulus**. Acervulus is saucer shaped with characteristic long and stiff setae. Conidia are single celled, hyaline, falcate. Conidiogenous cells are cylindrical phialides. The perfect state of *Colletotrichum* is *Glomerella*. **Ex :** *C. falcatum*	
(iii) Diplodia Pycnidium bears 2-celled, ovoid or ellipsoid conidia, dark brown in colour. Pycnidium globose, black and solitary. **Ex :** *Diplodia natalensis*	
(iv) Macrophomina Mycelium superficial or immersed, hyaline to brown in colour. Pycnidia globose, thick walled with central circular ostiole. Phialides determinate, obpyriform to cylindrical. Pycniospores are single celled, ellipsoid to obovoid. **Ex :** *M.phaseolina*	
(v) Phoma The conidial fructification is a **pycnidium**, a **dark-flask shaped** structure, opening usually by a single, circular ostiole and lined by a hymenium of conidiogenous cells from which numerous, single celled hyaline pycniospores develop. **Ex :** *Phoma lingam*	

Contd...

Colonial Morphology & Microscopic features	Diagram
(vi) *Septoria* Pycnidia are immersed in the host tissues and contain **filiform conidia** with several transverse septa. Perfect state is *Leptosphaeria*. **Ex :** *Septoria nodorum*	
(vii) Pestalotiopsis The genus can be easily recognized with its characteristic shape of conidia. Conidia are 5-celled with three central **cells coloured** and **two upper** and **lower cells being hyaline**, spindle shaped. The upper hyaline cell possess setulae and the lower cell gives rise to pedicel. **Ex :** *P. palmarum*	

Chapter 4
Nematophgous Fungi

Fungal parasites are important in natural regulations of plant parasitic nematode populations Nematophagous fungi are the fungi that feed on nematodes. These fungi can be classified into three types namely (a) fungal egg-parasites or opportunistic fungi, (b) nematode trapping fungi or predaceous fungi that capture nematodes using modified hyphal tips, (c) endoparasitic fungi that parasitize the nematode by means of small conidia or zoospores. There are various ways for soil-borne fungi to suppress nematode multiplication. In summary, there are four mechanisms that fungi use to suppress nematodes. Some of these interactions are direct where as others are indirect.

The direct mechanism is performed by fungi that feed on nematodes directly known as nematophagous fungi interact with nematodes in an indirect manner by several mechanisms including,

(i)   Fungi that kill nematodes by mycotoxins

(ii) Through the destruction of feeding sites of sedentary nematodes in roots.

(iii) Fungi that are non-pathogenic to plants, but complete with nematodes in roots and significantly reduce nematode multiplication. Many of these fungi are used as potential nematode biocontrol agents.

(iv) Mycorrhizal fungi improve the growth of nematode infected plants and may also affect the nematode development.

4.1 Occurrence

According to a survey of nematophagous fungi in Ireland by Gray (1983), nematophagous fungi were found in all the habitats examined, among which, permanent pasture, coniferous leaf litter, and coastal vegetation had the most frequent incidence of nematophagous fungi. Other habitats examined by Gray included coniferous leaf litter, old and partly revegetated dung, permanent grassland pasture, cultivated land, decaying vegetation and compost. In addition, many other studies also supported the hypothesis that nematophagous fungi are widely distributed and have great potential to be explored as biocontrol agents.

4.2 Fungi affecting Plant Parasitic Nematodes

Fungi may contribute upto 80% of the total microbial biomass in many soils. In nature, fungi continuously destroy nematodes in virtually all soils because of their constant association with nematodes in the rhizosphere. According to their feeding habits, nematophagous fungi are grouped into following categories.

4.2.1 Nematode Trapping Fungi (or) Predacious Fungi

These are facultative fungi that form trapping structures to trap nematodes. There are more than 50 species of predacious fungi which capture and kill nematodes in soils. For effective control, it is necessary that their limited period of activity coincides with the period of invasion of crop roots, as the age progresses, these fungi exhibit reduction in nematode–trapping efficiency. Predatory fungi produce extensive hyphal net work in the environment and produce trapping devices at intervals along each hyha and peak fungal activity is reached after 12-15 days of incubation. The traps produced by fungi capture nematodes either mechanically or by adhesion, allowing the fungus to penetrate the nematode rapidly and digest the contents.

Predatory fungi form six different types of traps such as

- **(i)   Adhesive hyphae**

    Ex : Zygomycotina, *Stylopage, Cystopage*

- **(ii)   Adhesive traps**

    Ex : Deuteromycotina

    *Monacrosporium cionopagum (*branches)

    M.elipsosporum (knobs)

    Arthrobotrys oligospora (net works)

- **(iii)   Non-adhesive traps**

    Ex : Deuteromycota

    Arthrobotrys dactyloides (Constricting ring)

    Dactylella leptospora (non-constricting ring).

4.2.2   Opportunistic Fungi

These are facultative parasitic fungi attacking sedentary stages of nematodes. Facultative fungi that are commonly soil saprophytes, and are opportunistic fungi isolated from sedentary stages such as female and egg stages of sedentary nematodes such as *Heterodera, Globodera* and *Meloidogyne.* They donot form specialized infection structures except appresoria. They can survive and proliferate in soil in the absence of nematodes. These fungi are capable of colonizing nematode reproductive structures. When once the fungi come in contact with egg masses and cysts, proliferate and grow rapidly and even parasitize all eggs that are in early embryonic development stages. It is generally believed that these fungi have better role in biocontrol of nematodes than other two groups.

Ex :   Hyphomycotina

Acremonium

Cylindrocarpon

Fusarium

Paecilomyces

Verticillium

4.2.3 Endoparasitic Fungi

These are obligate parasitic fungi that have limited growth in soil outside the colonized nematode cadaver.

(i)   they can infect vermiform nematodes by producing adhesive spores attached to the cuticle of passing nematodes.

   Ex :   Hyphomycotina

   Hirsutella rhossiliensis

   Drechmeria coniospora

   Verticillium spp.

(ii)   Some can infect vermiform nematodes by producing conidia spores that can be ingested by nematodes.

   Ex :   *Harposporium anguillulae*

(iii)   Some can infect vermiform nematodes by producing motile zoospores that encyst on the nematode's surface.

   Ex :   Oomycota

   Myzocytium spp.

   Lagenidium spp.

   Chytridiomycota

   Catenaria anguillulae

(iv)   Some can infect sedentary nematodes when the nematodes are exposed on the root surface

   Ex :   Oomycota

   Nematophthora gynophila

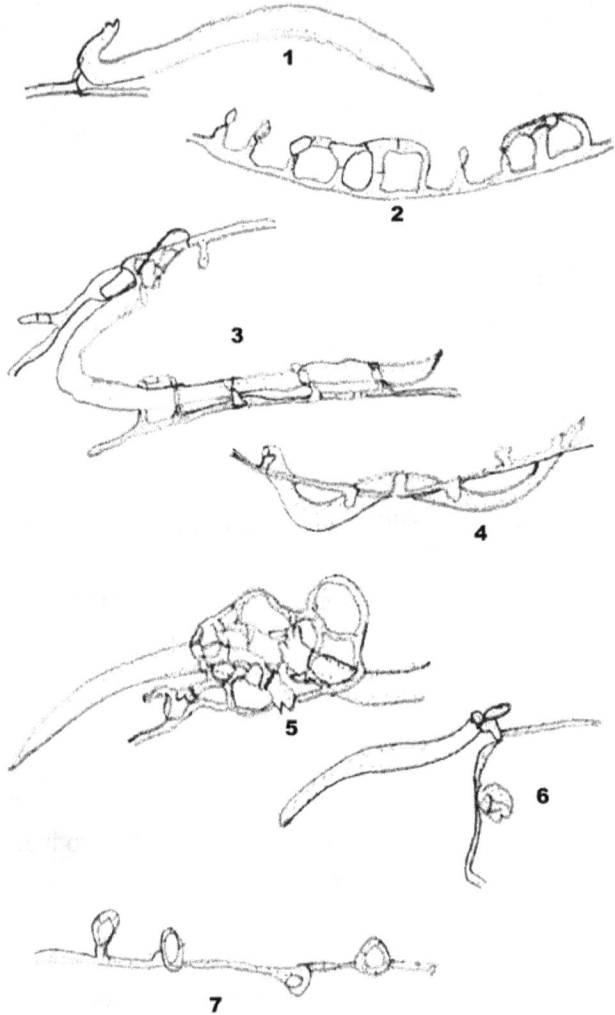

Fig. 4.1 Adhesive hyphae, branches, knobs, net and constricting rings.
1. Nematode attached to the adhesive hyphae of *Stylopage hadra* by
means of globular protuberance at the point of contact. 2. Hyphae of
Monacrosporium gephyrophagum producing adhesive columnar outgrowths
connecting with each other to from scalariform net work. 3. Nematodes
attached to adhesive branches of *M.cionopagum*. 4. Nematode captured by
adhesive knobs *M.parivcollis*. 5. Nematode entangled in adhesive network
of *Arthrobotrys musiformis*. 6. Nematode captured at head region in
constricting ring of *A.brochopaga*. 7. Open 3 celled constricting rigns of
A.brochopaga.

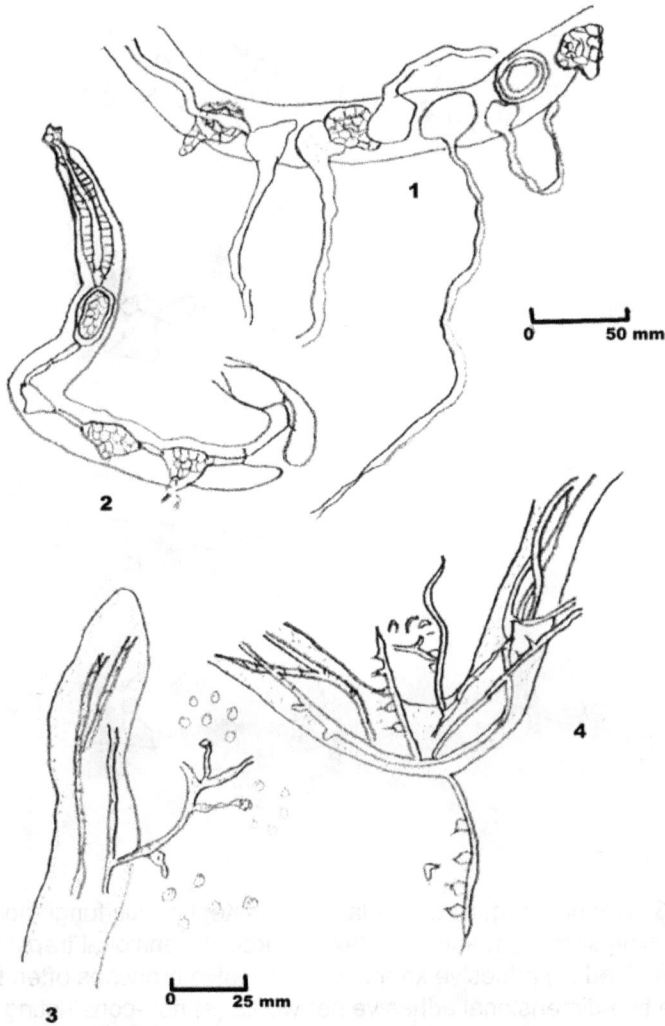

Fig. 4.2 Endoparasites. 1. Nematode body with infection hypha of *Myzocytium humicola*, empty sporangia having long, contortuous tubes and an oospore. 2. Nematode infected at the mouth with zoospores of *Catenaria anguillulae*. The whole body full of sporagia and two of them filled with zoospores. 3. Conidiophorous hyphae of *Acrostalagmus obovatus* with sterigmata and conidia. 4. Conidiophores of *Harposporium lillipulanum* emerging from an infected nematode, with phialides and conidia.

Fig. 4.3 Trapping organs of predatory nematophagous fungi: adhesive nets forming simple (a) and complex (b) three-dimensional traps; sessile (c) and stalked (d) adhesive knobs; and adhesive branches often forming simple two-dimensional adhesive net-works (e) non-constricting rings (g) and instructing rings open (h) and closed (i);

4.3   Key To Nematophagous Fungi
(Koon-Hui Wang and Robert Me Sorley, 2003)

This key for nematophagous fungi is simplified version from the key of Cooke and Godfreys (1963) with the focus an nematode-trapping fungi and endoparasitic fungi.

1.   Endoparasitic fungi (mycelium in the life cycle predominantly inside nematode host)

2.   Predatory fungi (mycelium in the life cycle predominantly outside nematode host)

3.   Assimilative hyphae within host transformed into fertile hyphae, extended out of host slightly, producing adhesive cells or ingestive conidia.

4.   Vegetative hyphae within host transformed into sporangia producing zoospores, or producing conidia, zygospores or azygospores

Endoparasitic fungi with adhesive cells or ingestive spores

3.   Hyphae aseptate ....

3.   Hyphae septate 4

4.   Hyphae with clamp connection 5

4.   Hyphae without clamp connection 7

    (a) *Nematoctonus*

5.   Hyphae bearing adhesive cells (knobs)

    (a) Nematoctonus robustus Jones

    (b) N.concurrens Drechs.

    (c) N.haptocladus Drechs.

    (d) N.campylosporus Drechs.

5.   Hyphae lacking adhesive cells, but producing adhesive knobs on conidium 6

6.   Chlamydospores produced

    (a) Nematoctonus pachysporus Drechs.

    (b) N.tylosporus Drechs.

6.   Chlamydospores not produced

    (a) Nematoctonus leiosporus Drechs.

    (b) N.leptosporus Drechs.

7.   Conidia borne on sterigmata, no phialides

    (a) *Drechmeria coniospora* (Drechs) Gams & Jansson

7.   Conidia borne on phialide.....................................8

8.   Conidia adhesive

    (a) Hirsutella rhossiliensis Miniter & Brandy

    (b) H.minnesotensis Chen, Liu, Chen

8.   Conidia filiform

    (a) Harposporium helicoides Drechs.

    (b) H.oxycoracum Drechs.

    (c) H.subuliforme Drechs

8.   Conidia arcuate

   (a) H.anguillulae Lohde (Karling)

   (b) H.liliputanum Dixon

   (c) H.crassum Shepard

8.   Conidia straight or slightly curved

   (a) H. baculiforme Drechs.

   (b) H.sicyodes Drechs.

8.   Conidia pea-pod, barbed at one or both ends

   (a) H.bysmatosporum Drechs.

   (b) H.diceraeum Drechs.

9.   Vegetative hyphae within the host developed into conidiophores that pass out of host, producing conidia.

   (a) Meristacrum asterospermum Drechs.

Endoparasites that produce Encysting Spores

9.   Vegetative hyphae within the host transformed into sporangia producing spores ...10

10.   Sporangium (zoosporangium) producing motile zoospores 11

10.   Sporagium producing immotile spores ....12.

11.   Zoospores uniflagellate, no zygospores, no resting spores.

   (a) Catenaria anguillulae sorokin

   (b) *Rhizophydium* sp.

11.   Zoospores biflagellate, may form zygospres, produce resting spores.

   (a) Lagenidium caudatum Barron

   *(b) Myzocytium vermicola (*Zopf) Fischer

   (c) M. glutinosporum Barron

   (d) M. humicola Barron & Percy

   (e) Nematophthora gynophila Kerrry & Crump

12.   Spores globular or polyhedral with a lobed appendages.

   *(a) Haptoglossa heterospora_*Drechs.

12.   Spores clavate.

   (a)   *Protascus subuliformis* Dangeard

Nematode – Trapping Fungi

13. Morphologically unmodified hyphae ....14.

13. Morphologically modified hyphae forming traps 17

14. Hyphae aseptate with yellow adhesive substances at contact ...15

14. Hyphae septate 16

Adhesive Mycelia

15. Produce conidia on simple conidiophore.

    (a) Stylopage hadra Drechs.

    (b) S.leiohypha Drechs.

    (c) S.grandis drechs

15. Without conidia, but chlamydospores formed.

    (a) Chlamydospores fomed laterally: *Cystopage lateralis* Drechs.

    (b) Chlamydospores formed intercalary: *C.intercalaris* Drechs.

    (c) Chlamydospores on crooked branches or intercalary : *C.cladospora* Drechs.

16. Conidia bifurcate

    (a) Triposporina aphanopaga Drechs.

16. Conidia furcated, trident-like.

    (a) T.ridentaria implicans Drechs.

17. Hyphae aseptate, lateral branches bearing poorly differentiated adhesive knobs.

    (a) *Acaulopage pectospora* Drechs.

17. Hyphae septate 18

18. Hyphae forming adhesive branches, sometimes forming simple 2-dimensional net work; conidiophore simple, single terminal conidium.

Adhesive Branches

    *(a) Monacrosporium cionopagum (*Drechs) Subram (Dactylella cionopaga Drechs)

    Synonym : *M.gephyropagum* (Drechs) Subram. *(Dactylella gephyropaga* Drechs.)

18. Hyphae forming stalked or sessile adhesive knobs 19

18. Hyphae forming stalked non-constricitng rings, sometimes accompanied by stalked adhesive knobs 21

18. Hyphae forming stalked constricting rings 22

18. Hyphae anastomosing to form 2 or 3 dimensional adhesive networks 23

Adhesive Knobs

19. Conidiophore branched

   (a) *Arthrobotrys haptospora (*Drechs.) Schenck, Kender & Pramer (*Dactylaria haptospora* Drechs)

   (b) *Monacrosporium haptotylum (*Drechs) Liu & Zhong (D.haptotyla Drechs)

   Synonym : *M.candidum* (Nees.) Liu & D.sclerohypha

   (c) *M.asthenopagum* (Drechs). Rdubner (Dactylella asthenopaga Drechs)

19. Conidiophore simple 20

20. Adhesive knobs always sessile

   (a) *Monacrosporium phymatopagum* (Drechs) Subram. (*Dactylella phymatopaga* Drechs)

20. Adhesive knobs sessile or short-stalked often form short chains of adhesive cells.

   (a) *M.parvicolle* (Drechs) Cooke & Dickinson (*Dactylella parvicollis* Drechs)

   (b) *M.lobatum* (Dudd.) Rubner (Dactylella lobata Dudd.)

   (c) *M.robustum* (McCulloch)

20. Adhesive knobs always stalked

   (a) *M.ellipsosporium* (Preuss) Cooke & Dickinson (Dactylella ellipsospora (Preuss) Grove)

   (b) *M.mammillatum* (Dixon) Cooke & Dickinson (*D.mammillata* Dixon)

Non-constricting Rings

21. Adhesive knobs not present.

   (a) *Monacrosporium leptosporum* (Drechs.) Rubner (*Dactylella leptospora* Drechs.)

21. Adhesive knobs present conidiophore simple.

   (a) *Monacrosporium lysipagum* *(*Drechs.) Subram. (*Dactylella lysipaga* Drechs.)

21. Adhesive knobs present conidiophore branched.

   (a) *Monacrosporium condidium* (Nees) Liu & Zhang (*Dactylaria candia* (Nees) Sacc. Drechs.)

   Synonym : *M.haptotylum* (Drechs) Liu & Zhang

Constricting Rings

22. Conidia borne in a terminal cluster on conidiophore.

   (a) *Arthrobotrys anchonia* Drechs.

   (b) *A.dactyloides* Drechs

   (c) *A. brochopaga* (Drechs.) Schenck, Kendrick & Pramer (*Dactylaria gracilis* Dudd)

22. Conidium borne singly on a simple conidiophore.

   (a) *Monacrosporium polybrochum* *(*Drechs.) Subram. (Trichothedcium polybrochum Drechs.)

   (b) *Monacrosporium acrochaetum* (Drechs) Cooke (*Dactylella acrochaeta* Drechs.)

   (c) *M.doedycoides* (Drechs.) Cooke & Dickinson (*D.doedycoides* Drechs.)

   (d) *M.stenobrochaum* (Drech.) Subram. (*D.stenobrocha* Drechs.)

   (e) *M.bembicodes* (Drech.) Subram (*D.bembicodes* drechs.)

   (f) *M.turkemenicum* (Sopronoy) Cooke & Dickinson (*D.Turkmenica* Sopronoy)

   (g) *M.coelobrochum* (Drechs) Subram. (*D.coelobrocha* (Drechs.)

3-dimensional Networks

23. Conidia with one septum

   (a) *Arthrobotrys cystoporia* (Dudd.) Mekht. (*Trichothecium cystoporium* Dudd.)

   (b) *Duddingtonia flagans* *(*Dudd.) Cooke (*T.flagrans* Dudd.)

(c)   *T.pravicovi* Soprunov

(d)   *T.globosporum* var *globosporum* Soprunov

(e)   *T.globosporum* var *microsporum* Soprunov

(f)   *T.globosporum* var rosem Soprunov

(g)   *Arthrobotrys arthrobotryoides* (Berl.) Lindau Drechs.

(h)   *A.conoides* drechs.

(i)   *A.oligospora* Fresenius

(j)   *A. superba* Corda.

(k)   *A.longispora* Soprunav

(l)   *A. oviformis* Soprunov

(m)   *A.doliformis* Soprunov

(n)   *A.Kirghizica* Soprunav

(o)   *A.cladodes var cladodes* Drechs.

(p)   *A.cladodes var macroides* Drechs.

(q)   *A. robusta* Dudd.

(r)   *A. musiformis* Drechs.

23.   Conidia with more than one septum.

(a)   *Monacrosporium   eudermatum*   (Drechs.)   Subram (Dactylaria eudermata Drechs.)

(b)   *M.   psychrophilum* (Drechs.) Cooke & Dickinson (Dactylaria psychrophila (Drechs) Subram.)

(c)   *M.   megglosporum* (Drechs) Subram. *Dactylella megalospora* Drechs.

(d)   *M. reticulatum* (peach) cooke & Dickinson (*Dactylella reticuluta* Peach)

(e)   *M.   thaumasium* (Drechs.) de Hoog & Oorschot (*Dactylaria thaumasia* Drechs.)

(f)   *Arthrobotrys polycephala* (Drechs.) Rifai (*D.polycephala* Drechs.)

(g)   *A. pyriformis* (Juniper) Schenck, Kendr. & Pramer (*Dactylaria pyriformis* Juniper)

(h)   *A. scaphoides* (Peach) Schenck, Kendr. & Pramer (*Dactylaria scaphoides* Peach)

(i)   *M.   gampsosporum* (Drechs.) Rubner (*Dactylaria gampsospora* Drechs.)

4.4 Improved Techniques for Detection of Nematophagous Fungi

There are several methods for isolation and observation of endoparasitic and nematode trapping fungi. With some applicable for quantitative or semi-quantitative evaluation of antagonistic activity. All the techniques however have certain disadvantages. Because the detection of slow developing endoparasites is often masked by fast growing trapping fungi, it is important to test for the two groups separately.

4.4.1 Baermann-funnel Technique (Nicolay and Sikora 1988)

The Baermann-funnel allows active nematodes passage through a filter specimen parasitized by endoparasitic fungi and not fully inactivated pass through the filter and parasitism is usually detected by transfer to agar plates. In this technique, the target nematodes are stained with chrysoidin. This stain was recommended by Doliwa (1955-56) and Kampfe (1956) as a viability stain. The technique allows direct detection of specific antagonistic of the nematode in question.

Procedure

- The Baermann-funnel is replaced by a dish that supports 100 g of soil which reduces variability obtained with other techniques using small amounts of soil.

- Field soil is placed on the dish on two milk filters over water and active nematodes for four days at room temperature.

- The extracted nematodes are discarded from the dish and 5000 nematodes of the target species in 10 ml of water spread evenly over the field soil on the filters.

Heavy soil should be dried for one or two days before the application of target nematode. Drying reduces exposure of endoparasitic fungi with adhesive conidia to excessive moisture that could reduce attachment to the nematode cuticle.

The dishes are not filled with water during the 48 hrs incubation stage. The nematodes are stained with chrysoidin for 24 hr prior to addition.

- The dishes are then filled with water for nematode extraction.

- 24 hr later, the nematode suspension in the dish is poured onto a 20 μm aperture sieve, washed with water and transferred into a 15 ml graduated centrifuge tube.

- The suspension is centrifuged for 5 min at 2500 rpm.

- The supernatant is removed with a pipette and the remaining 0.3 ml mixed with 0.3 ml solution containing 100 ppm streptomycin sulphate and 200 ppm penicillin.

- The resulting suspension is spread evenly over 1.5% water agar in a 9cm diam. Petridish.

- The dishes are kept at room temperature, examination for parasitized nematodes is made immediately after addition and at two day intervals for ten days.

4.4.2 Differential– Centrifugation Technique (Nicolay and Sikora, 1988).

The differential centrifugation technique, in use, seprates the heavier spores of the nematode trapping fungi from the lighter spores of endoparasites. The soil fraction containing the latter is poured onto a water agar petridish and spread with a glass rod over 50% of the dish surface. Nematodes are added in a water suspension. This suspension and the water from soil extract often spreads soil particles over the petridish surface making observations difficult.

In addition, fungi such as chytridiomycetes, oomycetes and zygomycetes are rarely detected. Furthermore, the amount of soil poured onto the agar surface is small which increases variation and requires the use of many replicates. There problems can be reduced by adding larger amounts of soil to grooves cut out of the agar. The sediment of the first centrifugation step that is routinely discarded can also be used for examination of nematode – trapping fungi.

Procedure

- 100 g of soil are suspended in 150 ml of water

- The suspension is poured onto a 250 μm aperture sieve and washed with as little water as possible through the sieve into a beaker.

- The filtrate is agitated for 10s, then poured onto a 100 μm aperture sieve and washed through the sieve into a beaker.

- The resulting suspension is centrifuged at 650g for 2 min.

- The supernatant is transferred to clean centrifuge tubes while the sediment containing the spores of nematode – trapping fungi is mixed with 1 ml of water, 1-1.5 ml of the soil suspension is pipetted into a

5-6 mm wide groove cut into the middle of a water agar petridish (9cm. dia.) by means of two joined scalpels.

- 3000 chrysoidin stained nematodes are added as boil in 0.6 ml of water.

- The supernatant of the first centrifugation step containing the spores of endoparasitic fungi is centrifuged again at 4060 g for 30 min.

- The sediment is resuspended and pipetted into the grooves in water agar. Endoparasitic fungi are best observed by cutting out a cross-shaped groove which increases nematode contact with fungal spores. In addition, this allows a 2 ml soil suspension to be used per plate.

- 3000 nematodes are pipetted onto the soil in the grooves for the detection of endoparasites. Nematode penetration into the agar is alleviated by adding a small amount of silver sand to the grooves.

- The plates are kept at room temperature with examination for parasitized specimens made immediately after addition and at two day intervals for two weeks.

The examination of trapping fungi requires more time and might take upto six weeks if examination is made for fungi with adhesive knobs or branches.

Advantages

- Detection of fugal antagonists in soil samples from the boil on a target nematode without special infectivity tests.

- They are efficient and time saving and give an overview of the spectrum of nematophagous fungi attacking mobile nematodes in field soil.

- Quantitative examinations are also possible.

- The stain allows marking of viable target nematodes, increases accuracy, and allows the use of a selected indicator species for specific fungal antagonism.

Fig. 4.4 Pure cultures and corresponding photomicrophs of nematophagous fungi. (from a to p).

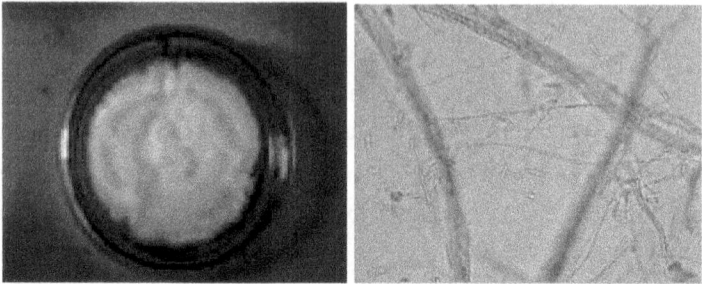

(a) *Fusarium sporotrichoides.*

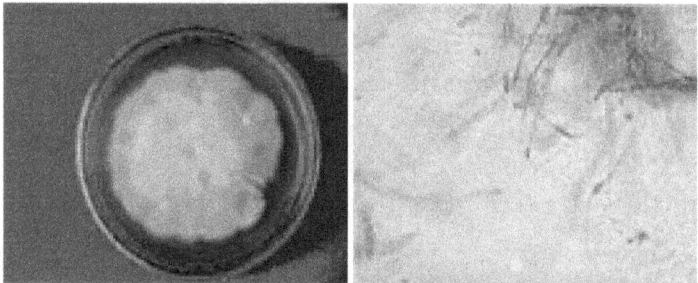

(b) *Fusarium oxysporum isolate*1.

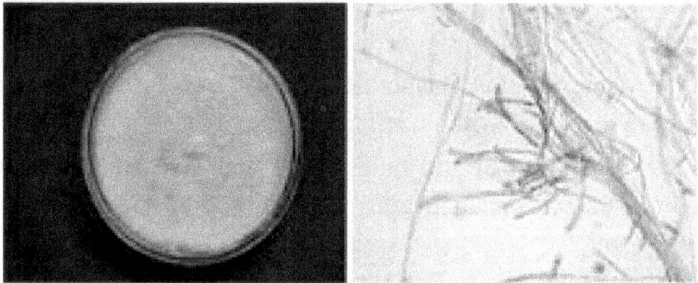

(c) *Fusarium oxysporum isolate* 2.

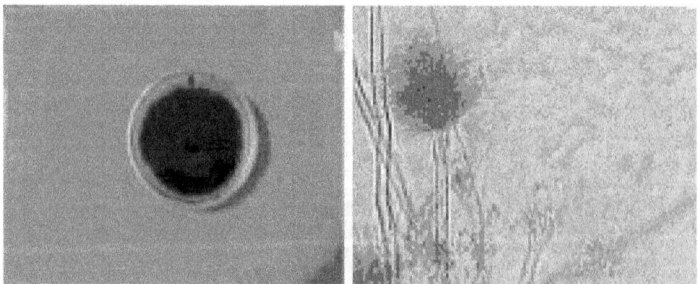

(d) *Aspergillus niger*

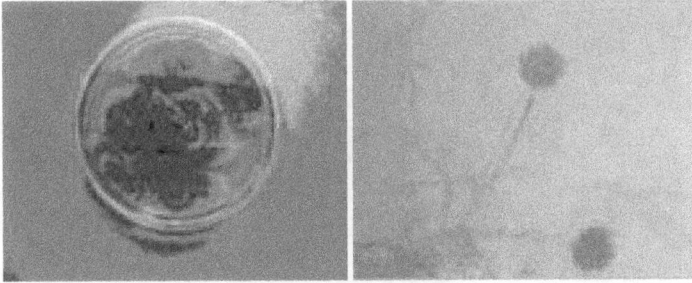

(e) *Aspergillus flavus.*

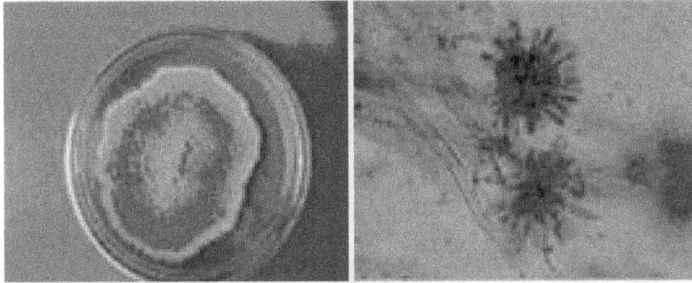

(f) *Aspergillus sp.*

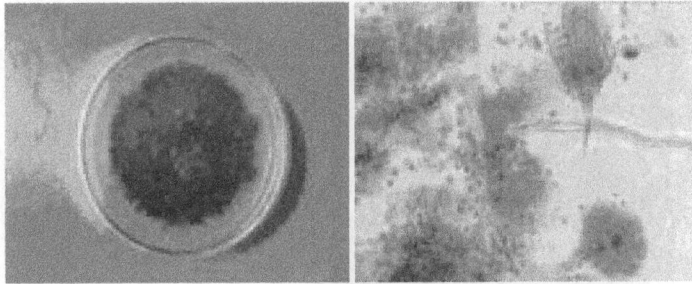

(g) *Penicillium sp.*

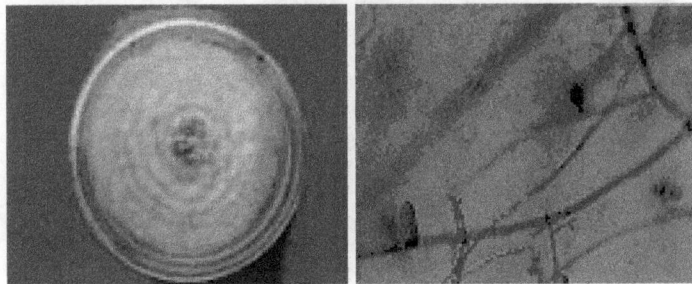

(h) *Helminthosporium sp.*

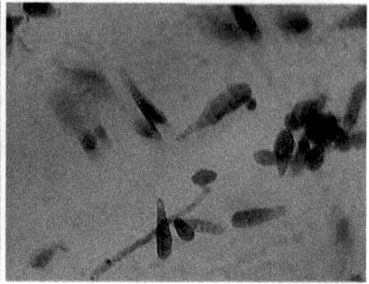

(i) *Alternaria sp.*

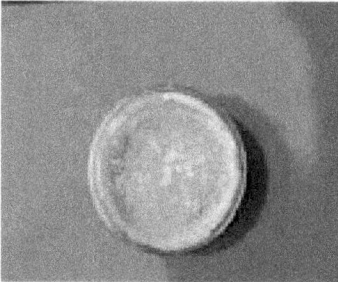

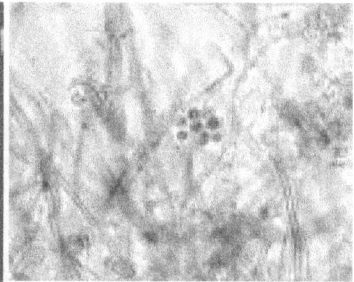

(j) *Botrytis sp.*

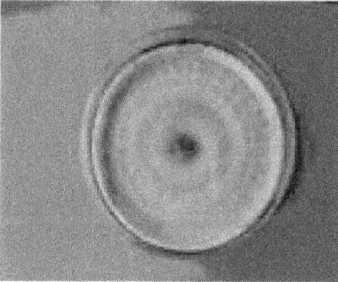

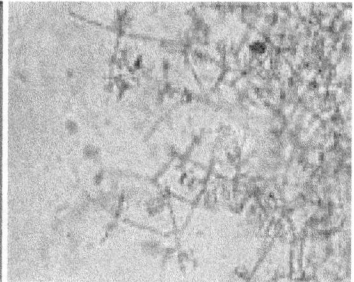

(k) *Trichoderma harzianum*

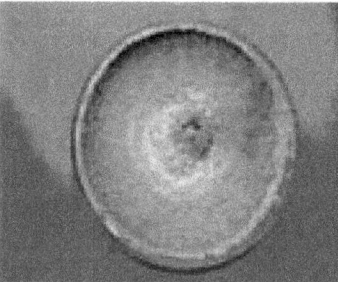

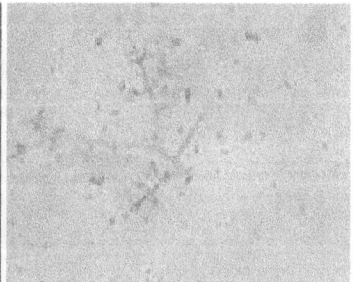

(l) *Trichoderma viride*

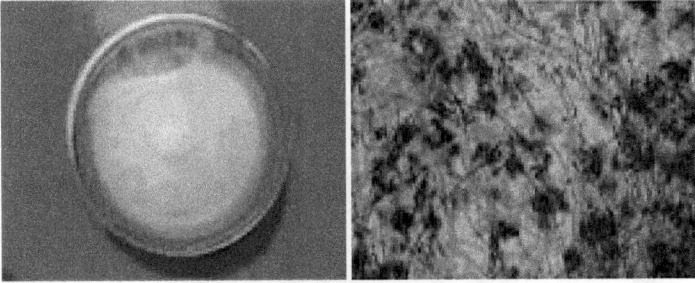
(m) *Paecilomyces lilacinus* isolate 1.

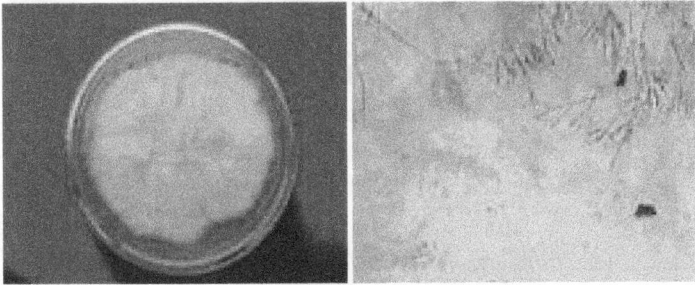
(n) *Paecilomyces lilacinus* isolate2.

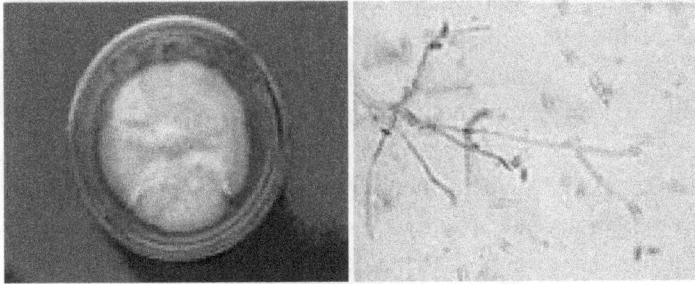
(o) Verticillium chlamydosporum isolate 1.

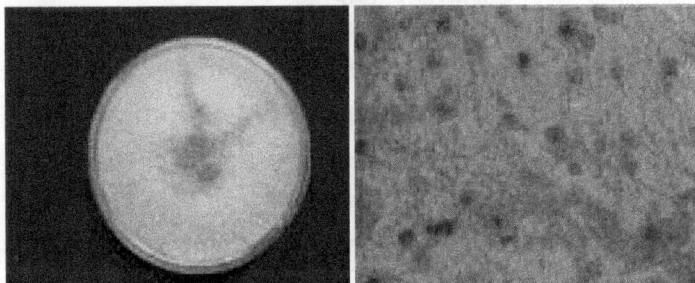
(p) *Verticillium chlamydosporum* isolate 2.

Chapter 5

Mycotoxins
Importance and their Detection

Mycotoxins are toxic metabolites produced by fungi, especially the saprophytic moulds growing on foodstuffs or animal feeds. These are the secondary metabolites elaborated by some toxic fungal strains, have received worldwide attention in recent years because of their manifestation with food materials as well as for their positive mutagenicity, carcinogenicity etc. Among the fungal genera, species of *Aspergillus, Penicillium, Fusarium* and *Alternaria* are the main producers of a wide range of mycotoxins. These fungi are well known for their capacity to spoil the associated substrates by degrading their nutritional components as well as by producing some important mycotoxins. The diseases caused due to consumption of mycotoxins is called 'Mycotoxicoses' have been responsible for major epidemics in man and animals during recent past. They must always been a hazard to human beings and domestic animals, but until the past 30 years their effects have been largely overlooked. Although poisonous mushrooms are carefully avoided, moulds growing as foods have been generally considered to cause unaesthetic spoilage, without being dangerous to health. Depending on the kind of toxin and dose, mycotoxins can be acutely chronically toxic.

5.1   Ill effects of Mycotoxins

- In animals, acute diseases include liver and kidney damage.
- Skin disorders, may be pecrotic lesions or photo sensitivity.
- Nerve toxins may cause trembling or even death.
- Hormonal imbalance include abortions in cattle, swollen genitals in pigs and a variety of poorly defined disorders including vomiting in pigs, refusal of feed.
- Central nervous system being affected.

Especially the toxins which act on liver and kidney are difficult to detect and levels much lower than those producing acute effects are often carcinogenic. When eaten in minute quantities in the daily diet, they can induce cancers in experimental animals long after the time of eating. It is probable that humans can be affected the same way.

5.2 Historical Background

Between 1960 and 1970, it was established that some fungal metabolites, now called mycotoxins were responsible for animal disease and death. In the decade following 1970, it became clear that mycotoxins have been the cause of human illness and death as well, and are still causing it. Among them, the most important had been ergotism, which killed thousands of people in Europe in the last thousand years.

- **1940** – Alimentary Toxic aleukia (ATA) – Responsible for death of thousands of people in USSR.
- **1930** : Stachybotryotoxicosis – which killed thousands of horses and cattle in USSR.
- **1960** : Aflatoxicosis – which killed 1 lakh young turkeys in England.

Mycotoxins are always hazardous to human and domestic animals. Besides cereals and pulses, fruits and vegetables are extensively consumed by human population due to their high nutritive values, flavor and taste. These substrates have also been shown to harbor a large number of microbes such as bacteria, fungi, viruses and actinomycetes.

Toxic effects : The effects of mycotoxins on animal and human health are multitude and with regard to human beings little information available because of the difficulty in carrying out research on human individuals. The toxicity of mycotoxins depends in particular as the character of molecule in question, the frequency of exposure and quantity absorbed. However, it should be remembered that doses administrated to animals during a study may be higher than the quantities normally present in food.

- **1974** – Aflatoxicoses – Moldy corn containing aflatoxin affected 15 villages in Gujarat and Rajasthan. Domestic animals consuming moldy corn also died. Analysis of samples indicated presence of aflatoxin levels 6.25 to 15.6 mg/kg corn which is far above the prescribed level.

5.3 Occurrence of Mycotoxins

Mycotoxin fungi are known to contaminate a variety of food materials as its production primarily depends upon nature of substrate and environmental conditions. Most of the tropical countries like India provide congenial climate for mycotoxin elaboration under natural conditions. Especially dry fruits will favour the growth of toxigenic fungi which ultimately leading to the elaboration of mycotoxins (Hanssen, 1971).

A number of dry fruits have been reported to be naturally contaminated with mycotoxin throughout the world. Among the mycotoxins, aflatoxins have been reported as the main contaminant of coconut (Singer, 1983) and other dry fruits viz., cashewnut, almonds, walnut, raisin, makhana and emblica (Narasimham, 1968). Other mycotoxins like zearalenone, citrinin and ochratoxin have also been recorded in some samples of dry fruits. Vegetables like tomato, brinjal, chillies and common fruits like orange, banana, apple, musambi and guava are also observed as good substrates for elaboration of mycotoxins under *in vitro* conditions. Even though aflatoxin contamination in fresh fruits and vegetables has not been investigated properly, it does not appear to be a serious problem for them because of their shorter storage span. However, presence of aflatoxins in some dried fruits, vegetables and rotting apples (Anon 1980) and in fruits like guava, Momordica, Charantia and Arocarpus lakoocha c sinha and Singh, 1982 b).

5.4   Types of Mycotoxins

Most commonly occurring mycotoxins are

- (a) Aflotoxins
- (b) Ochratoxins
- (c) Sterigmatocystin
- (d) Citrinin and Alternaria toxin
- (e) Zearalenone
- (f) Trichothecenes

5.4.1 Aflatoxins

These are produced primarily by *Aspergillus flavus* and *A. parasiticus.* Although two dozen varieties of aflatoxins are known such as A1, A2, B1, B2, G etc., only a few of them, most importantly aflatoxin A1, had been reported as naturally occurring compounds and the rest are secondary metabolites or derivatives (Bhat, 1991). Out of 18 isolates of *A. flavus* obtained from musambi, papaya, guava, potato, brinjal and Karela (*Momordica charantia* L.) 13 were found to be toxigenic and these produced aflatoxin B1 in the range of 0.07 to 1.72 ppm in liquid medium (Sinha & Singh, 1982 b). These toxigenic isolates were able to produce aflatoxins under natural conditions in guava, barhar and Karela fruits. Hansen and Jung (1973) recorded high incidence of *aflatoxin* producing *Aspergillus flavus* in various samples of walnuts, cashewnuts and almonds.

5.4.2 Zearalenone

It is produced chiefly by *Fusarium graminearum* however other strains of *Fusaria* such as *F.oxysporum*, *F.trichinctum* are also reported to synthesize it. This toxin is a known natural oestrogen which causes hormonal imbalance in certain species of animals, especially in pigs. Its effects on human beings are largely unknown. However, it is suspected of being the cause of precocious onset of puberty in Puerto-Rican children.

5.4.3 Trichothecenes

These toxins can be divided into two types viz., simple and macrolyclic. The simple trichothecenes are mainly produced by *Fusarium graminearum* and *F. tricinctum* while the macrocyclic trichothecenes are produced by *Myrothecium verrucaria*, *Trichothecium roseum* and *Stachybotrys*.

Examples for simple Trichothecenes : T-2 toxin, Diacetoxy scirpenol (DAS), Deoxynivalenol (DON), Verrucarol

Macrolyclic Trichothelenes : Roridin R, Verrucarin A, Strato toxin H

5.4.4   Ochratoxins

It constitutes a group of structurally related metabolites that are produced by *Aspergillus ochraceus* as well as *Penicillium viridicatum, P. cyclopium, P.expansum* and *P.palitans*. The major mycotoxin in this group is ochratoxin A (OTA). A few isolates of A. *flavus*, P. *citrinum* and *Fusarium moniliforme* were found to be associated with those samples in which high incidence of toxigenic *A. flavus* was recorded in cashewnut, almond, coconut, markhana, embilica and walnut (Bilgrami, 1985).

5.4.5 Sterigmatocystin

It is produced by several species of *Aspergillus*. Among them, *A. versico* for being the major producer. Chemically sterigmatocystin resembles aflatoxin and it is a precursor in the biosynthesis of aflatoxin (Hsieh, et al., 1973).

5.4.6 Citrinin and Alternaria Toxins

Citrinin, a yellow coloured mycotoxin is produced chiefly by *Penicillium citrinum, P.expansum* and *P.verrucosum* (Frisvad, 1987). It is a significant renal toxin to monogastric domestic animals including dog (Carlton et al 1974). *Penicillium expansum* isolated from tomato produces patalin and citrinin. Alternaria toxins include a variety of toxic secondary metabolites produced by various species of *Alternaria* (Blaney, 1991 Watson, 1984). Five of the Alternaria toxins viz., alternariaol, alternariol methyl ether, tenuazonic acid, alternuene and alter toxin-I have been identified as common food contaminants (Visconti, 1986). Species of *Alternaria* are also of common occurrence on fruits and vegetables as many of them are plant pathogens that damage crops in the field or cause post-harvest decay. Mislivec et. al., (1987) reported toxigenic strains of *A. alternata* and *A.tomato* associated with fresh tomatoes used ketchup production. The important toxic metabolites produced by them are alternariol, alternariol methyl ether, Tenuazonic acid, altertoxin I, II and III.

5.5   Tolerence Limit of Various Mycotoxins

5.5.1 Vomitoxin (DON)

Cattle / Chickens	:	10 ppm
Swine	:	5 ppm
All other animals	:	5 ppm
Human food	:	1 ppm

5.5.2 Fumanisins

Rabbits	:	5 ppm
Swine	:	20 ppm
All other animals	:	10 ppm
Human food	:	2 ppm.

5.5.3 Aflatoxins

Breeding cattle, Swine And mature poultry	:	100 ppb
Finishing Swine	:	200 ppb
Finishing beet cattle	:	300 ppb
Human food	:	20 ppb

5.6   Quantification and Detection of Mycotoxins

For detecting the presence of mycotoxins there are mainly three types of assays have been available.

(a)   Biological methods

(b)   Analytical methods

(c)   Immunological methods.

Biological assays are used when analytical and immunological methods are not available for routine analysis. Biological assays are qualitative and are often non-specific and time consuming. For estimation of mycotoxins, many analytical and immunological methods are available. Various analytical methods viz., Thin layer chromotogrphy (TLC), High performance liquid chromatography (HPLC). Pyrolysis mass spectrometry have been developed for mycotoxins analysis. With the availability of monoclonal and polyclonal antibodies against mycotoxins, various simple, sensitive and specific ELISA tests are developed for analysis of mycotoxins.

5.7 Chromatography

Chromatography is one of the most useful and popular analytical technique dealing with the separation of closely related compounds from a mixture *i.e.,* amino acids, proteins, lipids, simple sugars, etc. The credit for the discovery of chromatography goes to the Russian botanists Mikhail Tswett who coined the term chromatography (chroma – colour, graphein – to written). Chromatography means colour writing and it is the process of separating substances by the development of chromatograms. Chromatograms are the collection of different bands on the chromatographic column.

Chromatography is the process which permits the resolution of mixtures by effecting separation of all or some of their components in concentrated zones. In this process, the individual pigments are removed from the adsorbing material by a suitable solvent. Adsorbents used are Talc, alkaline earth oxides, CO_2, PO_4, starch and cane sugar, silicagel.

In chromatography – two phases
- Stationary – Solid, liquid or gel
- Mobile or moving – Liquid or gaseous

Types
- Thin layer chromatography
- Paper chromatography
- Ion exchange chromatography
- High performance liquid chromatography (HPLC)

5.7.1 Thin layer Chromatography

A thin layer of stationary phase is spread on glass plate. The movement of the mobile phase across the layer generally by simple capillary action. As the mobile phase moves across the layer from one edge to the opposite, it transfers the analytes placed on the layer. The movement of analytes is expressed by retardation factor (or) relative front (R_F).

$$RF = \frac{\text{Distance moved by analyte from origin } (d_A)}{\text{Distance moved by solvent front from starting point } (d_m)}$$

The efficiency of a thin layer is expressed by number of theoretical plates, N where

$$N = 16 \left(\frac{d_A}{W} \right)^2$$

W = Width of the analyte

$$\text{Plant height, } H = \frac{d_A}{N}$$

The capacity factor 'K' for the analyte is $\quad K = \dfrac{d_m}{d_A}$

Separation of sugars – Thin layer chromatography (Silica gel)

Materials

- Thin layer of Silica gel
- Solvent – Ethyl acetate : isopropanol : water : pyridine (26 : 14 : 17 : 2)
- Standard sugar solution – Lactose, ribose, Rhamonase, Xylose, Fructose, glucose
- Mixture of unknown sugars
- Aniline – diphenylamine reagent

Method

- Carefully spot the sugars on the plate with a 5 μl pipette without making a hole in the adsorbent.
- Place the plate in a chamber saturated with solvent and develop the chromatogram.
- Draw a line across the plate at the point of chromatogram.
- Remove the chromatogram when the solvent reaches the mark.
- Dry the plate in a stream of cold air.
- Locate the sugars by spraying with Aniline – diphenylamine reagent.

Advantages of TLC

- Noticeably sharper separations can be obtained.
- A higher sensitivity.
- Requires only limited sample quantities with a minimum of approximately 0.5 mg.

- A large number of samples can be run simultaneously in less time.

- Much faster development than paper chromatography.

5.7.2   High Performance Liquid Chromatography (HPLC)

Chromatographic methods are used for the determination of mycotoxin content in cereals and other agricultural commodities, especially high performance liquid chromatography and gas chromatography methods are widely used for routine analysis. These methods are mainly used for the final seperation of matrix components and detection of analyte of interest in reliability, these methods are widespread compared to thin layer chromatography (TLC). However, TLC remains the method of choice for rapid screening practices and for situations where advanced HPLC equipment is not available.

In HPLC, a liquid mobile phase (or solvent) is used to transport the sample through the column, which is packed with an immobilized liquid stationary phase. The analyte is then partitioned between the two phases as it makes its way through the column, leading to a seperation of compounds due to different partitioning coefficients. There are two types of HPLC namely normal phase chromatography with a polar stationary phase (eg. Silica gel) and a non-polar solvent (eg. Hexane) or vice-versa, using a C8 or C18- hydrocarbon phase 3with a polar solvent such as water, methanol or acetonitrile. This approach is called reversed phase chromatography (RP-HPLC) and is commonly used in mycotoxin analysis.

5.7.3   Liquid Chromatography with Mass Spectrometric Detection (LC – MS) and Biosensors

Liquid chromatography with mass spectrometric detection (LC-MS) and biosensors for mycotoxins represent fairly recent developments in mycotoxin determination. On the other hand, biosensors are subject to ongoing research in order to provide suitable and easy to use tools for the determination of fungal infection and toxins. Despite high costs and the need for experienced personnel, LC – MS techniques are already finding widespread use in mycotoxin analysis. Extraction and clean-up techniques have to be applied prior to seperation and detection in order to enable well separated peaks without interference from matrix components.

5.7.4   Detection of Aflatoxins

The presence of aflatoxins in the infected sample can be detected by using thin layer chromatography (TLC) method.

The contaminated seeds or sampling material is surface sterilized by using sodium hypochlorite 1.0% or chlorine 0.4%. The fungi associated with the infected material can be isolated using a selective DG-18 medium and incubate the plate near UV light under 12 hr. light and 12 hr darkness for 7 days at 25 °C. The different fungal species are identified using stereo binocular microscope. For isolation of different species of Aspergillus, Oat meal agar (OMA) medium is used. Czapekdox medium is used for maintenance, by toxins 0.5 ml of molten agar + 0.05% detergent in a screw cap vial.

Preparation of TLC plates : Remove all grease marks and finger prints present on the plate using acetone. Silica gel is used as supporting medium. 30 gm of silica gel mixed with 60 ml of distilled water and shake it mechanically for few minutes and spread the slurry with the help of plate leveler and see that thickness of coating should be 0.25 cm. Then dry the plates keeping it in a dust free chamber. After proper drying, plates are exposed to high temperature at 110^0C in oven and cool it using Agar plug method, cut the mycelial disc of Aspergillus sps. And wet the mycelial plug with chloroform / methanol (2:1) ratio, Ensure that mycelial side of wet agar plug should touch the TLC plate and diameter of the application should not exceed 0.6 cm and let the spots dry and there should be 17-23 plays and standard develop the TLC plate using eluent TEF solvent and view the plate in day light and at different wavelengths (short wave length 254 nm and long wave length 366 nm).

5.8   List of Mycotoxins

• Acetoxyscirpenediol	• Lateritin +1
• Acetyldeoxynivalenol	• Luteoskyrin
• Acetylneosolaniol	• Lycomarasmin +1
• Acetyl T-2 toxin	• Lysergic Acid
• Aflatoxin	• Macrocyclic Trichothecenes
• Aflatrem	• Malformin
• Altenuene	• Maltoryzine
• Altenuic Acid	• Moniliformin
• Altenusin	• Monoacetoxyscirpenol
• Alternariol	• Neosolaniol

Contd...

• Altertoxin	• Niidulotoxin
• Austdiol	• Nitropropionic Acid
• Austamide	• Nivalenol
• Austin	• NT-1 Toxin
• Austocystin	• NT-2 Toxin
• Avenacein +	• Ochratoxin
• Beauvericin +2	• Oxalic Acid
• Bentenolide	• Oxaline
• Brefeldin	• Patulin
• Brevianamide	• Penicillic Acid
• Butenolide	• Penicillic Acid
• Calonectrin	• Rhizonion
• Chaetosin	• Roridin A.E
• Chaetoglobosin	• Rosetoxin
• Chetomin	• Roquefortine
• Citrinin	• Rubratoxin
• Citreoviridin	• Rubroskyrin
• Citromycetin	• Rubrosulphin
• Cladosporic Acid	• Rugulosin
• Cochliodinol	• Rugulovasine
• Crotocin	• Sambucynin + 1
• Cyclopiazonic Acid	• Satratoxins, F.G.H
• Cyclosporin A	• Scirpentriol
• Cytochalasin E	• Secalonic Acid
• Deacetylcalonectrin	• Slaframine
• Deoxynivalenol Diacetate	• Sporidesmin
• Deoxynivalenol Monoacetate	• Sterigmatocystin
• Diacetoxyscripenol	• T-1 Toxin
• Diplodiatoxin+	• T-2 Toxin
• Destruxin B	• Tenuazonic Acid
• Egrine	• Territrem
• Emodin	• Tremorgenic
• Enniatins	• Triacetoxyscirpendiol
• Ergometrine	• Trichodermin

Contd...

• Ergonovine	• Trichodermol
• Ergotamine	• Trichothecenes
• Ergotoxine	• Trichothecin
• Erythroskyrin	• Trichoverrins
• Fructigenin + 1	• Trichoverrols
• Fumagilin	• Tryptoquivalene
• Fumitoxin	• Viridicatin
• Fumitremorgen	• Verrucarin
• Fumonisin B1	• Verruculogen
• Furanocoumarins	• Verrucosidin
• Fusaric Acid	• Viopurpurin
• Fusarin	• Viomellein
• Fusarochromanone	• Vioxanthin
• Gliotoxin	• Viriditoxin
• Griseofulvin	• Walleminol
• HT-2 Toxin	• Xanthocillin
• Ipomeanine	• Xanthomegnin
• Islanditoxin	• Yavanicin + 1
• Isosatratoxin	• Zearalenone
• Koninginin	•

5.9   Some Common Mycotoxins and the Organisms that Produce Them

Mycotoxin	*Organism*
Acetoxyscirpenediol	*Fusarium monilliforme, F.equiseti, F.oxysporum, F.culmorum, F.avenaceum, F.roseum, and F.nivale*
Acetyldeoxynivalenol	*Fusarium monilliforme, F.equiseti, F.oxysporum, F.culmorum, F.avenaceum, F.roseum, and F.nivale*
Acetylneosolaniol	*Fusarium monilliforme, F.equiseti, F.oxysporum, F.culmorum, F.avenaceum, F.roseum, and F.nivale*
Acetyl T-2 toxin	*Fusarium monilliforme, F.equiseti, F.oxysporum, F.culmorum, F.avenaceum, F.roseum, and F.nivale*
Afatoxin	*Aspergillus flavus, A.parasiticus*

Table Contd...

Mycotoxin	Organism
Aflatrem	*Aspergillus flavus*
Altenuic acid	*Alternaria alternate*
Alternariol	*Altenaria alternate*
Austdiol	*Aspergillus ustus*
Austamide	*Aspergillus ustus*
Austocystin	*Aspergillus ustus*
Avenacein + 1	*Fusarium monilliforme, F.equiseti, F.oxysporum, F.culmorum, F.avenaceum, F.roseum, and F.nivale*
Beauvericin + 2	*Fusarium monilliforme, F.equiseti, F.oxysporum, F.culmorum, F.avenaceum, F.roseum, and F.nivale*
Bentenolide	*Monographella nivalis*
Brevaianamide	*Aspergillus ustus*
Butenolide	*Fusarium monilliforme, F.equiseti, F.oxysporum, F.culmorum, F.avenaceum, F.roseum, and F.nivale*
Calonectrin	*Fusarium monilliforme, F.equiseti, F.oxysporum, F.culmorum, F.avenaceum, F.roseum, and F.nivale*
Chaetoglobosin	*Chaetomium globosum*
Citrinin	*Aspergillus carneus, A.terreus, Penicillium citrinum, P.hirsutum, P.verrucosum.*
Citreoviridin	*Aspergillus terreus, Penicillium citreoviride*
Cochliodinol	*Chaetomium cochliodes*
Crotocin	*Acremonium crotocinigenum*
Cytochalasin E	*Aspergillus clavatus*
Cyclopiazonic acid	*Aspergillus versicolor*
Deacetylcalonectrin	*Fusarium monilliforme, F.equiseti, F.oxysporum, F.culmorum, F.avenaceum, F.roseum, and F.nivale*
Deoxynivalenol diacetate	*Fusarium moniliforme, and F.nivale*
Deoxynivalenol monoacetate	*Fusarium monilliforme, F.equiseti, F.oxysporum, F.culmorum, F.avenaceum, F.roseum, and F.nivale*
Diacetoxyscirpenol	*Fusarium moniliforme, F.equiseti*
Destuxin B	*Aspergillus ochraceus*

Table Contd...

Mycotoxin	*Organism*
Enniatins	*Fusarium monilliforme, F.equiseti, F.oxysporum, F.culmorum, F.avenaceum, F.roseum, and F.nivale*
Fructigenin + 1	*Fusarium monilliforme, F.equiseti, F.oxysporum, F.culmorum, F.avenaceum, F.roseum, and F.nivale*
Fumagilin	*Aspergillus fumigatus*
Fumonisin B1	*Fusarium monilliforme, F.equiseti, F.oxysporum, F.culmorum, F.avenaceum, F.roseum, and F.nivale*
Fusaric acid	*Fusarium moniliforme*
Gliotoxin	*Alternaria, Aspergillus fumigatus, Pencillium*
HT-2 toxin	*Fusarium monilliforme, F.equiseti, F.oxysporum, F.culmorum, F.avenaceum, F.roseum, and F.nivale*
Ipomeanine	*Fusarium monilliforme, F.equiseti, F.oxysporum, F.culmorum, F.avenaceum, F.roseum, and F.nivale*
Islanditoxin	*Pendicillium islandicum*
Lateritin + 1	*Fusarium monilliforme, F.equiseti, F.oxysporum, F.culmorum, F.avenaceum, F.roseum, and F.nivale*
Lycomarasmin + 1	*Fusarium moniliforme*
Malformin	*Aspergillus niger*
Maltoryzine	*Aspergillus niger*
Moniliformin	*Fusarium monilliforme, F.equiseti, F.oxysporum, F.culmorum, F.avenaceum, F.roseum, and F.nivale*
Monoacetoxyscirpenol	*Fusarium monilliforme, F.equiseti, F.oxysporum, F.culmorum, F.avenaceum, F.roseum, and F.nivale*
Neosolaniol	*Fusarium monilliforme, F.equiseti, F.oxysporum, F.culmorum, F.avenaceum, F.roseum, and F.nivale*
Nivalenon	*Fusarium monilliforme, F.equiseti, F.oxysporum, F.culmorum, F.avenaceum, F.roseum, and F.nivale*

Table Contd...

Mycotoxin	Organism
NT-1 toxin	*Fusarium monilliforme, F.equiseti, F.oxysporum, F.culmorum, F.avenaceum, F.roseum, and F.nivale*
NT-2 toxin	*Fusarium monilliforme, F.equiseti, F.oxysporum, F.culmorum, F.avenaceum, F.roseum, and F.nivale*
Ochratoxin	*Aspergillus ochraceus, Penicillium viridictum*
Oxalic acid	*Aspergillus niger*
Patulin	*Aspergillus clavatus, Pencillium expansum, Botrytis, P.roquefortii, P.claviforme, P.griseofulvum,*
Penicillic acid	*Aspergillus ochraceus*
Penitrem	*Penicillium crustosum*
Roridin E	*Myrothecium roridum, M.verrucaria, Dendrodochium spp., Cylindrocarpon spp. Stachybotrys spp.*
Rubratoxin	*Penicillium rubrum*
Rubroskyrin	*Pencillium spp.*
Rubrosulphin	*Pencillium viridicatum*
Rugulosin	*Penicillium brunneum, P.kloeckeri, P.rugulosum*
Sambucynin + 1	*Fusarium monilliforme, F.equiseti, F.oxysporum, F.culmorum, F.avenaceum, F.roseum, and F.nivale*
Satratoxins, F.G.H.	*Stachybotrys chartarum, Trichoderma viridi*
Scirpentriol	*Fusarium monilliforme, F.equiseti, F.oxysporum, F.culmorum, F.avenaceum, F.roseum, and F.nivale*
Slaframine	*Rhizoctonial leguminicola*
Sterigmatocystin	*Aspergillus flavus, A.nidulans, A.versicolor, Penicillium rugulosum*
T-1 toxin	*Fusarium monilliforme, F.equiseti, F.oxysporum, F.culmorum, F.avenaceum, F.roseum, and F.nivale*
T-2 toxin	*Fusarium monilliforme, F.equiseti, F.oxysporum, F.culmorum, F.avenaceum, F.roseum, and F.nivale*

Table Contd...

Mycotoxin	*Organism*
Triacetoxyscirpendiol	*Fusarium monilliforme, F.equiseti, F.oxysporum, F.culmorum, F.avenaceum, F.roseum, and F.nivale*
Trichodermin	*Trichoderma viride*
Trichothecin	*Trichothecium roseum*
Trichoverrins	*Stachybotrys chartarum*
Trichoverrols	*Stachybotrys chartarum*
Tryptoquivalene	*Aspergillus clavatus*
Verrucarin	*Myrothecium verrucaria, Dendrodochium spp. Stachybotrys chartarum*
Verruculogen	*Aspergillus fumigatus, Stachybotrys chartarum*
Viopurpurin	*Trichophyton spp. Penicillium viridicatum*
Viomellein	*Aspergillus spp., Pencillium aurantiogriseum, P.crustosum, P.viridicatum*
Viriditoxin	*Aspergillius fumigatus*
Xanthocillin	*Erotium chevalieri*
Yavanicin + 1	*Fusarium monilliforme, F.equiseti, F.oxysporum, F.culmorum, F.avenaceum, F.roseum, and F.nivale*
Zearalenone	*Fusarium monilliforme, F.equiseti, F.oxysporum, F.culmorum, F.avenaceum, F.roseum, and F.nivale*

Chapter 6
Isolation and Inoculation of Bacteria

The correct diagnosis of any disease is a prerequisite of control. The more rapidly and accurately the causal organism is identified, the sooner can proper controls be instituted. Although precise chemotherapy can be applied to most human bacterial infections once the pathogen is identified, specific chemical treatment for controlling plant pathogenic bacteria is in an early stage of development. Antibiotics and various formulations of copper are used to control plant bacterial diseases but with limited success.

Apathy toward research on ecology and control of bacterial diseases is due in large part to difficulties in identifying the pathogens. There is a real need for development of effective and rapid methods for identification of plant pathogenic bacteria. Much of this increased need for rapid identification is a result of a much increased international trade and movement of plant propagative materials such as potato tubers and seeds. Biochemical and physiological tests which are routinely used to identify plant pathogenic bacteria are not entirely satisfactory. The tests are often complicated, difficult to interpret, and require weeks to months to complete. Also, not all strains of each organism always give the same results.

The use of serology for identifying bacteria is almost as old as the science of Plant Pathology itself. Many medically important bacteria are routinely identified by serological tests. The ability of medical bacteriologists to provide quick and accurate serological identification of bacteria indicates a similar potential for identification of plant pathogenic bacteria.

6.1 Isolation of Bacterial Plant Pathogens

Selection of infected material occupies a key role for successful isolation of bacterial pathogens. The bacteria can be readily isolated from young lesions usually but in advanced stages isolation of bacteria is difficult due to overhauled population of saprophytic microorganisms which would over cover the pathogen in isolation plates.

In cases if the disease specimen does not contain young lesion, it is always preferable to inoculate suspension of the diseased tissue on the healthy host plant and the lesions that are developed can be used for isolation. Before subjecting to isolation, it is ideal to go for ooze test and staining to be performed for the presence of bacterium.

6.2   Suspension Preparation

The selected diseased lesion should be washed with ethyl alcohol. Then dip in mercuric chloride 0.1% for 15 seconds and pass it through three changes of sterile distilled water. Dip a slide in spirit, flame it through spirit lamp and allow to cool it. Place few drops of water on the slide and put the surface sterilized lesion on the slide. If the affected tissue show abundant oozing, the lesion is cut into two halves with a sharp blade and permit the diffusion of bacteria into water. Where as if the tissue showing feable bacterial ooze, it should be teased apart with a sterilized blade to get the bacteria into suspension. The prepared suspension can be utilized for isolation of bacteria by following purification methods.

6.2.1 Streak Plate Method

In this method, pour about 20 ml of molten nutrient agar medium (2% agar) in a petriplate cooled to about 45°C and then allow the medium to solidify. After few hours (usually 2-3 hrs), these plates are used for inoculation purpose. Take a loopful of suspension prepared by maceration of diseased tissue, streak over the surface of agar medium in zig- zag fashion. Two more plates are inoculated without recharging the wire loop with bacterial suspension. Incubate the plates in an incubator at 25 °C in inverted position and observe for colony growth daily. Most of phytopathogenic bacteria develop colonies within 4-5 days. More number of discrete and individual colonies are obtained in 2nd or 3rd plate.

6.2.2 Pour Plate Method

In this method twenty ml of nutrient agar medium is dispensed in to three test tubes. Then, inoculate one with a transfor it to second tube and loopful of bacterial suspension prepared from diseased tissue and mix thoroughly by rotating tube holding between palms. Take a loopful of mixture from first test tube and mix it thoroughly. Takeout a loopful of mixture from second test tube and transfer it to third tube and mix thoroughly. Then, pour the above three dilutions into three petriplates separately. Allow the plates to solidify and keep it in incubator in an inverted position after labeling and observe for colony growth.

Composition of Nutrient Agar medium

Beef Extract	:	5 gm
Peptone	:	10 gm
Agar	:	20 gm
Distilled water	:	1000 ml
pH	:	7-0

It is often possible that only the colonies of the pathogen will develop with well chosen diseased tissue. If more than one type of colonies are formed, it is preferable to select those which are abundant and consistently found in different suspensions of the diseased tissue. Usually, the slow growth colonies are likely to be that of pathogenic bacteria and colonies which usually appear before 7 hrs of inoculation are not likely to be pathogenic. Sometimes, it is desirable to select two or three types of colonies. The one which proves pathogenicity can be selected and remain are discarded after multiplication on nutrient agar medium.

When the right type of colony is chosen, transfer it onto nutrient agar slants. Take small inoculum by touching the wire loop on well isolated colony and inoculate into a agar slant. The cultures which are obtained by single colony transfer, are further subjected to purity. Prepare a dilute suspension of culture in water and streak the suspension on nutrient agar plates. In case of pure culture, only one type of colonies formed with original characteristics.

6.3 Inoculation of Bacteria

When a culture is used for testing the pathogenicity, care should be taken that same cultivar is used for testing from where the culture is isolated. Young and vigorously growing plants are to be selected for inoculation. The plants are to be kept in humid chamber before and after inoculation. Whichever may be the inoculation method, control plants should be maintained where the plants are treated with distilled water. The following inoculation methods are used based on type of symptoms.

6.3.1 Root-dip method

In case of soil-borne diseases like vascular wilts, this method seems to be the most natural way of inoculating bacterial pathogen. Uproot the young seedlings, remove soil particles adhere to the roots by running under tap water and immerse in bacterial suspension after clipping the tips of roots and then transplant the seedlings.

6.3.2 Pin-prick Inoculation

Prepare a pin-bundle by fixing 4-6 fine insect pins on a piece of cork. Care should be taken that only the tips of pins should project out of the cork piece. Pinch the plant part to be inoculated with pin-bundle and bacterial suspension is applied on the injured part with a cotton swab. This method is good for bacterial pathogens such as blight, wilt and soft rot. In the case of blights, the leaf is inoculated in the centre barring the mid-rib region. Where as in wilts, inoculation is to be carried in the region between cotyledon and first leaf.

6.3.3 Injection or Leaf Infiltration

In this method the bacterial suspension is injected into intercellular spaces of leaves with a hypodermic needle (25 gauge) and pin size should be 5/8. The hypodermic needle is inserted gently under the epidermis of the leaf and it should be observed that opening face of needle should move towards the leaf. The desired amount of inoculum is injected with the help of hypodermic syringe so that leaf tissue becomes water soaked. This method is commonly used for inoculating stem and other parts of plant.

6.3.4 Spray Inoculation

This method of inoculation is most common in case of diseases like leaf spot, blight and canker where the pathogen usually enters the host through stomata, hydathodes or lenticels. Keep the plants in moist chamber for 2, 4 hr prior to inoculation to facilitate the stomata to open and to create high intercellular humidity in the tissues around natural openings. Spray the bacterial suspension with hand atomizer to cover the entire plant surface thoroughly. After spraying, keep the plants in moist chamber for 48 hrs again and observe for symptoms.

6.3.5 Spray Inoculation under Pressure

This method reduces the time taken for symptom development and is highly suitable for bacterial pathogens those who invade stomata. In this case, the bacterial suspension is forced through the stomata into intercellular spaces. For obtaining related type of symptoms, diluted bacterial suspension is preferable. Where as, use of concentrated suspension results in typical symptoms. The inoculum is applied with an atomizer at 1.5 kg/cm^2 pressure on the lower surface of leaf. The leaf is held in position with hand and nozel of atomizer should be kept at about two inch distance. Pressure is created by connecting the atomizer to the exhaust outlet of electric motor pump.

6.3.6 Smear Inoculation

This is best applicable in cases of plant parts with waxy coat or high surface tension and which do not allow the sprayed inoculum to stick onto the surface of plants. The inoculum is smeared on the surface of the plant with the help of muslin cloth soaked in the suspension. After application, keep the plants in humid chamber or spray with water thrice a day until symptoms are developed.

For the above methods the inoculum is prepared by suspending a 24-48 hr agar slant culture in about 25-30 ml of sterile distilled water.

After inoculation, the plants are to be observed for symptom development, at least for 30 days. In case of diseases like galls caused by *Agrobacterium tumafaciens* for symptom development, it may take more than a month. Certain leaf spots such as mango canker may take as long as 10-15 days. Where as wilts and blights take 7-10 days to develop symptoms. It is important to record the stage of the development along with description of symptoms.

If the symptoms produced by the inoculated bacterium resemble the original specimen from where it was isolated, is considered as a true pathogen. Plant pathogenic bacteria produce different disease symptoms on most plant species whether they are typical or atypical. If the symptoms are replicate in artificially inoculated plant as that of original one and symptom development is slow, are called typical. Where as in atypical symptoms rapid necrosis of tissue results without any water soaking at the site of inoculation. Use of high inoculum concentration may result in development of atypical symptoms, particularly when inoculation is done by pin-prick, injection infiltration, though the bacterium may be phytopathogenic. These atypical symptoms are the resultant of hypersensitive reaction and should not be misconstrued for the true pathogenic reaction.

Chapter 7
Detection of Bacterial Infection

For detection of bacterium in infected tissue, ooze test is performed to confirm their presence. It is desirable to see the presence of bacterium before proceeding to isolation.

7.1 Ooze Test

For performing the test, take a clean glass slide, place few drops of water onto it. Then, cut a piece through infected tissue with a sharp razor blade and place it on glass slide containing water drops and put coverslip. Examine the slide under compound microscope for the presence of ooze.

A cloudy mass of bacterial cells is seen oozing out through cut ends of the infected tissue if the infection occurs due to bacteria. The test also provide an idea regarding vascular or parenchymatous infection. Bacteria oozes out forcefully at distinct points corresponding to vascular strands in case of vascular infection. Where as in parenchymatous infection, bacteria oozeout slowly and it is diffused throughout the cutends. But in case of certain diseases like hairy root, leafy gall and crown gall some difficulty arises where all the diseased portion may not contain bacterial cells, and in tissues which contain large amount of starch, dispersal of starch grains in water will mask the bacterial ooze and further makes it difficult to recognize. But, with some experience it is possible to differentiate bacterial ooze from starch content.

7.2 Preparation of Stained Smears

In some cases, the affected tissue contain very low bacterial infection and distinct bacterial ooze may not be detected in such cases Ex : Red stripe of sugar cane and mung bean leaf spot caused by Xanthomonads. To reveal the presence of bacterial cells, staining of smear is performed.

For this purpose, take a portion of infected tissue, place it on one end of slide in few drops water and crush it with a razor blade and allow to stand for one minute. Then tilt slowly, in such a way that water should flow towards the other end leaving the crushed host tissue. Prepare smear out of it, dry the smear by passing over a Bunsen flame. Flood the smear with crystal violet or carbol fuchsin and let it for one minute. Then, drain the stain and wash it thoroughly in running tap water. Blot dry the slide and examine under oil immersion objective for the presence of bacterial cells.

If the bacterium is not detected by both ooze test and also by staining, using of such material for isolation of bacteria goes waste.

7.3 Determination of Location of Pathogen

In sections of fresh material, there is chance of losing bacterial population when sections are transferred to water from the razor. Hence, it is preferable that the material is fixed in a suitable killing and fixing solution in order to determine the exact location of bacteria in diseased tissue. The most commonly used killing and fixing solutions are formalin -acetic acid – alcohol (FAA) and FAA saturated with mercuric chloride. The tenacity with which bacterial cells bind together and keeps the bacterial mass without dispersal during sectioning and staining.

Composition of FAA

Formalin	-	6.5 ml
Acetic acid	-	2.5 ml
Alcohol	-	100 ml

Fixation of infected material

Cut the diseased material into 5mm bits, and transfer in FAA solution. The minimum time required for killing and fixing is 48 hrs, but the material can be kept for any length of time. In FAA solution saturated with mercuric chloride, the tissue pieces are kept for 48 hr and then washed in FAA to remove excess of mercuric chloride and then preserved in FAA solution.

7.4 Staining of Diseased Sections

The infected leaf material could be cut with a sharp razor blade and staining can be performed using acid fuchsin or by differential staining.

7.4.1 Staining with Acid Fuchsin

The bacterial cells can be seen prominently in the affected tissues of leaf sheath, stem, coleoptile etc contain little or no chlorophyll by this method. Mount the section cutting in 0.1% acid fuchsin in lactophenol and allow to stand for 5-10 minutes and examine under microscope. In microscopic examination, it reveals that the bacteria are stained deep red and host cell walls appear pink in colour.

7.4.2 Differential Staining

By this method, the bacteria stain blue and the host cell walls appear yellow or green and the liquefied tissue appear light blue.

First, stain the infected material with Carbol thionin for about 5 minutes (Thionin blue 0.1 gm, phenol crystals 5.0 gm, water 100ml). Remove the excess stain by washing in water and apply ethyl alcohol 95%. Differential stain such as orange – G is applied. Keep it for few minutes. After few minutes, wash in absolute alcohol. Clear in xylol and mount in balsam.

Chapter 8

Characterization of Phytopathogenic Bacteria

The culture which proves pathogenic to plant is further to subjected to be studied for its morphological, cultural and physiological characters. Morphological characters are studied using Gram staining, Flagella staining, Capsule staining, etc.

8.1   Gram Staining

The most important and widely used staining procedure in Bacteriology is Gram staining. It is also called as differential staining as more than one stain is involved in the reaction. The most important differential stain used in bacteriology is the Gram stain, named after Dr. Christian Gram. It divides bacterial cells into two major groups, gram-positive and gram-negative, which makes it an essential tool for classification and differentiation of microorganisms. The Gram stain uses four different reagent. Description of these reagents and their mechanisms of action follow.

Differential staining requires the use of at least three chemical reagents that are applied sequentially to a heat-fixed smear. The first reagent is called the primary stain. Its function is to impart color to all cells. In order to establish a color contrast, the second reagent used is the decolorizing agent. Based on the chemical composition of cellular components, the decolorizing agent may or may not remove the primary stain from the entire cell or only from certain cell structures. The final reagent, the counterstain, has a contrasting color to that of the primary stain. Following decolorization, if the primary stain is not washed out, the counterstain cannot be absorbed and the cell or its components will retain the color of the primary stain. If the primary stain is removed, the decolorized cellular components will accept and assume the contrasting color of the counterstain. In this way, cell types or their structures can be distinguished from each other on the basis of the stain that is retained.

Crystal violet is used as a primary stain. This violet stain is used first and stains all cells purple. Gram's Iodine reagent serves as a mordant, a substance that forms an insoluble complex by binding to the primary stain. The resultant crystal violet-iodine (CV-1) complex serves to intensify the color of the stain, and all the cells will appear purple-black at this point. In gram-positive cells only, this CV-1 complex binds to the magnesium-ribonucleic acid component of the cell wall. The resultant magnesium-ribonucleic acid-crystal violet-iodine (Mg-RNA-CV-1) complex is more difficult to remove than the smaller CV-1 complex. Ethyl alcohol, 95% is used as a decolourising agent. This reagent serves dual function as a lipid solvent and as a protein dehydrating agent. Its action is determined by the lipid concentration of the microbial cell walls. Safranin is used as a counter stain. This is the final reagent, used to stain red to those cells that have been previously decolorized. Since only gram-negative cells undergo decolorization, they may now absorb the counterstain. Gram-positive cells retain the purple color of the primary stain.

Procedure

Prepare a smear of the organism and flood the smear with crystal violet and let it stand for one minute. Remove excess stain running under tap water. Flood the smear with Gram's Iodine and allow to stand for 1 minute and wash with tap water. Decolorize with ethyl alcohol (95%). Add reagent drop by drop till no crystal violet comes from the smear. Wash with tap water and apply counter stain, safranin and allow to stand for 45 seconds.

Using sterile technique, prepare a smear of each of the organism. Remove excess stain running under tap water. Blot dry and examine under oil immersion.

8.2   Capsule Staining

The capsule staining is performed to distinguish the capsular material from the bacterial cell. A capsule is a gelatinous outer layer that is secreted by the cell and that surrounds and adheres to the cell wall. It is not common to all organisms. Cells that have a heavy capsule are generally virulent and capable of producing disease, since this structure protects bacteria against the normal phagocytic activities of host cells. Chemically, the capsular material is a polysaccharide, a glycoprotein, or a polypeptide.

Capsule staining is more difficult than other types of differential staining procedures because the capsular materials are water-soluble and may be dislodged and removed with vigorous washing. Smears should not be heated, because the resultant cell shrinkage may create a clear zone around the organism that is an artifact, that can be mistaken for the capsule.

Crystal violet is used as a primary stain. A violet stain is applied to a non-heat-fixed smear. At this point, the cell and the capsular material will take on the dark color.

Copper sulfate (20%) can act as both decolorizing agent and counter stain because the capsule is nonionic, unlike the bacterial cell. The primary stain adheres to the capsule without binding to it. Since the capsule is water-soluble, copper sulfate, rather than water, is used to wash the purple primary stain out of the capsular material without removing the stain that is bound to the cell wall. At the same time, it acts as a counterstain as it is absorbed into the decolorized capsular material. The capsule will now appear light blue in contrast to the deep purple color of the cell.

Procedure

Take out grease free glass slides and prepare a heavy smear of the organism, allow the smear to air dry and be cautious that it should not be heat fixed. Flood the smear with crystal violet and allow to stand for about 5 to 7 minutes. Wash the smear with copper sulphate solution (20%). Gently blot dry and observe under oil immersion (100 X).

8.3 Spore Staining (Schaeffer-Fulton Method)

Spore staining is performed for differentiating bacterial spore with that of vegetative cell forms. Members of the anaerobic genera *Clostridium* and *Desulfotomaculum* and the aerobic genus *Bacillus* are examples of organisms that have the capacity to exist either as metabolically active vegetative cells or as highly resistant, metabolically inactive cell types called spores. When environmental conditions become unfavourable for continuing vegetative cellular activities, particularly with the exhaustion of a nutritional carbon source, these cells have the capacity to undergo sporogenesis and give rise to a new intracellular structure called the **endospore**, which is surrounded by impervious layers called spore coats. As conditions continue to worsen, the endospore is released from the degenerating vegetative cell and becomes an independent cell called a spore. Because of the chemical composition of spore layers, the spore is resistant to the deleterious effects of excessive heat, freezing, radiation, desiccation, and chemical agents, as well as to the commonly employed microbiological stains. With the return of favourable environmental conditions, the free spore may revert to a metabolically active and less resistant vegetative cell through germination. It should be emphasized that sporogenesis and germination are not means of reproduction but merely mechanisms that ensure cell survival under all environmental conditions. The spore stain uses two different reagents.

Malachite green is used as a primary stain. Unlike most vegetative cell types that stain by common procedures, the spore, because of its impervious coats, will not accept the primary stain easily. To further penetration, the application of heat is required. After primary stain is applied and the smear is heated, both the vegetative cell and spore will appear green. Once the spore accepts the malachite green, it cannot be decolorized by tap water, which removes only the excess primary stain. The spore will remain green. On the other hand, the stain does not demonstrate a strong affinity for vegetative cell components; the water removes it, and these cells will be colorless. Safranin is used as a counter stain. This contrasting red stain is used as the second reagent to color the decolorized vegetative cells, which will absorb the counterstain and appear red. The spores retain the green of the primary stain.

Procedure

Prepare individual smears in the usual manner using sterile technique, allow the smear to air dry and heat fix it. Flood the smear with malachite green and place on a hot plate, allow the preparation to steam for 2 to 3 minutes. Care should be taken that evaporation of stain is not allowed, if needed replenish the stain. Remove slides from hot plate, cool it and wash under running tap water. Apply counter stain, safranin and let it stand for 30 seconds and wash with tap water. Blot dry and observe under oil immersion.

8.4   Flagella staining

Bacterial cells possess flagella as external organelles which are responsible for motility. Flagella are very delicate and fragile, therefore the cultures are handled very carefully if these organs are to be stained.

The width of these organs is smaller than the wavelength of light and thus they cannot be seen without staining or with ordinary staining. During staining, the flagella is to get a heavy deposition of stain. For this purpose, mordants like tannic acid, potassium aluminium sulphate and mercuric chloride are used. So that, width of flagella will be more than the wave length and can be seen under microscope.

The glass slides used for preparing smears for flagella staining should be free of grease. For cleaning the slides make vim powder paste in water and apply it on both surfaces of the slide and allow to dry it partially. Remove the coating with a cotton swab with gentle rubbing. Like this the process is repeated until no trace of vim available. To test the slides free from grease, keep the slide in slanting position and put a drop of water on the upper end. If the drop slowly flows down to the lower end of the slide leaving a uniform water film along its track indicates that slides are free of grease and can be readily used.

Preparation and Fixation of Smear

Most of the phytopathogenic bacteria produce copious amount of polysaccharides when bacteria are grown on sugar containing media. Further, it is evident from the literature that when bacteria are grown in media containing high sugars may not develop flagella. For this, use 12-18 hr old culture grown on slants of nutrient agar medium which does not contain sugar. The sterile distilled water is poured gently into the culture tube without disturbing the bacterial growth and it should cover entire bacterial growth. Keep it undisturbed for about 2-3 minutes and let all the motile cells come into suspension and make the water turbid. Slightly agitate the tube, if the bacterial cells do not come into suspension and again let it undisturbed for two minutes.

Place a drop of water on the upper end of the slide and keep it in slanting position and allow the drop to run down and air dry at room temperature. Mark the smeared area on the opposite side of the slide with a marker pen.

(i) Gray's Method

Staining solutions

Solution A

Mercuric chloride	-	2.0 ml
Tannic acid		
(20% aqueous solution)	-	2.0 ml
Aluminium Potassium Sulphate	-	5.0 ml
(Saturated aqueous solution)		

Solution B

Basic Fuchsin	-	0.4 ml
(Saturated alcoholic solution)		

Mix solution A and B the solutions can be separately preserved for a long time. But, it will deteriorate after mixing and hence, they should be used immediately after mixing.

(ii)   Ziehl – Neelsons Carbol Fuchsin method

Basic fuchsin	-	0.3 gm
Phenol crystals	-	5.0 gm
Ethyl alcohol (95%)	-	700 ml
Sterile distilled water	-	95.0 ml

Dissolve basic fuchsin in alcohol and dissolve phenol in water. Then, mix the two solutions.

Procedure

- Prepare smear on cleaned glass slides as described earlier.
- Apply the mordant solution onto the slide until the smear is covered and allow to react it for 10 minutes.
- Wash the slide in distilled water.
- Filter Ziehl Neelsons Carbol Fuchsin on the slide to cover the smear and allow to stain for 5 minutes.
- Remove the excess stain in running tap water.
- Blot it dry and examine under oil immersion objective.

After staining, examine the whole smear and select a microscopic field showing maximum number of bacterial cells with flagella. Usually the motile cells tend to accumulate at the margins of the smear and therefore, it is preferable to observe bacterial cells, at the periphery of margins.

Chapter 9
Cultural Characteristics of Bacteria

To observe colony morphology of bacteria, prepare a dilute suspension of a 48 hr. old culture and streak the suspension on the nutrient agar medium and incubate the plates at 25°C in an inverted position for the development of single colonies. The characters studied are size, shape of bacterial cells, texture, surface marking, elevation, colour, time of appearance etc. Production of pigments or crystals should also be observed.

9.1   Temperature

Most of phytopathogenic bacteria are mesophiles i.e. they grow well between a temperature range of 25-30°C, except in some cases like *Pseudmonas solanacearum* which requires an optimum temperature of 35-37°C and certain *Clavibacterium* species have a temperature range of 20-25°C. Thus, temperature plays an important role in the growth of bacteria and based on temperature relationship certain *Pseudomonas* species are differentiated.

For this purpose, inoculate the test bacterium in several nutrient agar slants and keep it for incubation at temperatures ranging between 1-45°C. After proper incubation record the time of appearance of visible growth and amount of growth visually. The optimum temperature of the bacterium is determined by noting the temperature at which maximum amount of visible growth appears. Whereas at temperatures, where culture fails to grow is considered as maximum and minimum temperatures.

9.2   Oxygen Requirement

To determine whether a culture is aerobic or not, prepare nutrient dextrose agar containing 0.005% Bromocresol purple columns in tubes. Inoculate the tubes in duplicate by thrusting the inoculation needle straight into the medium containing bacterial suspension. Then, cover the agar surface in one tube with sterile parafin oil to a depth of 1 cm and incubate the tubes at 25°C. Record observations regularly at 48 hrs interval for about 7 days. After the growth of organism, if the colour of the medium changes from blue to yellow, it is a facultative anaerobe whereas, the growth of the organism and colour of the medium changes only in one tube with parafin and the tube without paraffin would indicate anaerobic and aerobic nature of the organisms respectively. Most of the plant pathogenic bacteria are aerobic in nature. But, some of the genera like *Erwinia* and *Pseudomonas* species grow anaerobically.

9.3   Production of Levan

Production of levan is detected using the following medium.

Beef extract	:	0.5 gm
Peptone	:	10.0 gm
Sucrose	:	50.0 gm
Agar	:	20.0 gm
Distilled water	:	1000 ml.

Sterilization of medium is done by subjecting to steaming for 30 min. on 3 successive days. Pour the medium into sterile petriplates. Streak a dilute suspension of the test culture on solidified agar medium and incubate the teplates at 25°C. Production of large, white and mucoid colonies indicate levan. Most of the phytopathogenic pseudomonads are characterized producing levan.

9.4   Pigment Production

Phytopathogenic bacteria produce pigments are referred to as chromogenic. Two types of pigments are known to be produced by bacteria such as water soluble and water insoluble pigments. Pigments which can diffuse into the medium are referred as water soluble. While, pigments which do not diffuse into the medium said to be water insoluble.

Yeast glucose chalk agar medium is used for the production of water insoluble pigments. Composition of the medium is as follows.

Chalk ($CaCo_3$)	:	20 gm
Glucose	:	10 gm
Yeast extract	:	10 gm
Agar	:	20 gm
Distilled water	:	1000 ml.

Take 500 ml beaker and dissolve yeast extract and glucose in 400 ml of distilled water. Take another beaker and dissolve agar in 200 ml of water. Mix the two solutions and make upto 1000 ml. Slants are prepared by dispensing the medium into tubes and sterilized by autoclaving. Inoculate the culture on the slants and observe the colour of the pigment regularly after 48 hrs.

King's medium is used for detecting the production of water soluble fluorescent pigment

K$_2$HPO$_4$	:	1.5 gm
Mg SO$_4$.7H$_2$O	:	1.5 gm
Peptone	:	20.0 gm
Glycerine	:	10.0 ml
Agar	:	20.0 gm
Distilled water	:	1000 ml
pH	:	7.2

The medium is prepared, autoclaved and poured into sterile petriplates. The culture is streaked on the plate in zig-zag fashion. Plates are incubated at 25°C and observe for pigment formation after 48 hrs. If a diffusible pigment is produced by the organism, a clear pigment zone is produced around the growth, which can be detected easily.

Chapter 10

Physiological Characteristics of Bacteria

To characterize phytopathogenic bacteria, studies based on physiological characters would form an important criteria. The Erwinias and Pseudomonands even upto species level are differentiated based on physiological characters. But in case of Xanthomonands, they are useful to study only upto genus. Differentiation of species is not possible because of the intraspecies variation.

10.1. Fermentation of Carbohydrates and Related Compounds

The ability to ferment carbohydrates would be considerably vary with different bacterial species. As some of the bacteria invade many carbohydrates while, others may not ferment any. Some bacteria are able to produce both acid and gas while fermenting carbohydrates. Whereas in some cases only acid is produced.

There is no method of assessing the fermenting ability of the bacterium beforehand. For this, tests are to be conducted. If an organism ferments any carbohydrate means it will be most likely glucose. If glucose is being fermented, then other monosaccharides can be tested before proceeding to carbohydrates of greater complexity.

The following carbon compounds are employed in the test

Pentoses	:	Ribose, Xylose, Arabinose, Rhamnose
Hexoses	:	Glucose, Galactose, Fructose, Mannose
Trisaccharides	:	Rattinose, Melezitose.
Polysaccharides	:	Cellulose, Starch, Glycogen, Cellulose, Inulin.
Sugar alcohols	:	Arbitol, Glycerol, Mannitol
Non-carbohydrate	:	Inositol

A synthetic medium is used as a basal medium for testing carbohydrate fermentation and organic nitrogen sources such as peptone or aminoacids should be avoided as far as possible. If bacterium fails to grow in inorganic nitrogen sources, organic nitrogen sources may be added in small concentrations. As most of the plant pathogenic bacteria are weak acid produces alkalinity produced during utilization of organic sources of nitrogen may neutralize the acidity.

(a) *Ayers, Rupp and Johnson's medium* – The medium can be used for most of the phytopathogenic bacterium excepting some species of *Pseudomonas* and *Xanthomonas*. The finished medium should be reddish – violet in colour.

KCl	-	0.2 gm
$(NH_4) H_2PO_4$	-	1.0 gm
$Mg SO_4. 7H_2O$	-	0.2 gm
Agar	-	20 gm
Bromocresol purple	-	0.7 ml
Distilled water	-	1000 ml

Basal medium for **Xanthomonads (Dye, 1962)**

K_2HPO_4	-	0.5 gm
$MgSO_4, 7H_2O$	-	0.2 gm
$NH_4H_2PO_4$	-	0.5 gm
NaCl	-	5.0 gm
Yeast extract	-	1.0 gm
Agar	-	20.0gm
Distilled water	-	1000 ml
pH	-	6.8

(b) **Basal medium given by Shaffer (1967)**

NH_4Cl	-	1.0 gm
K_2HPO_4	-	4.0 gm
KH_2PO_4	-	4.0 gm
$Mg SO_4.7 H_2O$)	-	0.2 gm
$Fe SO_4$	-	0.02 gm
$Zn SO_4$	-	0.003 gm
Nicotinic acid	-	0.01 gm
Agar	-	20.0 gm
Phenol Red	-	0.018 gm
Distilled water	-	1000 ml

To the above basal media, carbon compound to be tested is added so as to give 1.0% concentration.

10.2 Starch Hydrolysis

Starch is a polymer of glucose units joined through alpha – glucoside linkages. The hydrolysis of starch occur when the bacterium produces an extra cellular enzyme amylase. Amylase, an extra cellular enzyme which hydrolyses starch. Soluble type of starch is employed in this test. The enzyme alpha-amylase cause partial hydrolysis to erythrodextrin and dextrin. Whereas, Beta – amylase hydrolyses starch completely to maltose units. For detecting the hydrolysis of starch, Iodine solution is used. It gives a blue colour with starch, brown colour with erythrodextrins and no colour with maltose.

Starch agar medium is used for hydrolysis and composition as follows.

Beef extract	-	5.0 gm
Peptone	-	10.0 gm
Starch	-	2.0 gm
Agar	-	20.0 gm
Distilled water	-	1000 ml
pH	-	7.0

Sterilize Starch Agar (SA) medium by subjecting to autoclave and pour into sterilized petriplates. Let the medium solidify and spot inoculate the test culture in 4 plates. Four different cultures can be inoculated into a plate and incubate the plates at 25°C and test for starch hydrolysis, one plate at a time, after 2,4,7 and 14 days after inoculation. Flood the agar surface with Lugol's iodine and allow to react for a few minutes. If starch is hydrolysed, a colourless zone is observed around the bacterial growth in contrast to the blue background of the medium.

10.3 Gelatin Liquefaction

Gelatin is a protein, which is insoluble in cold water but soluble in hot water and on cooling, it forms a transparent gel. It is extracted from skin, ligaments, bones etc in boiling water. It is being attacked and decomposed by many bacteria which produce an extracellular enzyme namely gelatinase. As a result of decomposition, it loses the property of gel formation. These are two methods available to test gelatin liquefaction.

10.3.1   Stab Method

In this method, Gelatin medium is used.

Composition

Peptone	-	10.0 gm
Beef extract	-	5.0 gm
Gelatin	-	20.0 gm
Distilled water	-	1000 m
pH	-	7.0

Procedure

Mix all the ingredients together and heat over a water bath until the gelatin is dissolved. Dispense the medium in tubes, cool the tubes and allow to stand at 20^0C for 2 days to check the sterility. Inoculate these gelatin columns by stabbing a straight inoculation needle charged with 48 hr growth of the test bacterium. Incubate the tubes at 20^0C. Observe for liquefaction of the gel column at regular intervals upto one month. The form in which the gel is liquefied is also noted.

10.3.2   Gelatin-agar Medium

Composition

Peptone	-	10.0 gm
Beef extract	-	3.0 gm
Gelatin	-	4.0 gm
Agar	-	20.0 gm
Distilled water	-	1000 ml
pH	-	7.0

Procedure

The medium is sterilized in flasks by autoclaving and cooled to 45°C. It is then poured into petriplates and allowed to solidify. The medium is spot inoculated by a 48 hr growth of the test bacterium. Four cultures can be inoculated in a plate. After incubation of 48 hrs. to 7 days at 25°C, flood the agar surface with 0.2% mercuric chloride solution in dilute hydrochloric acid (20%) and allow to react for a few minutes. The reagent forms a white precipitate with gelatin. A clear zone occurs, if the bacterium has liquefied gelatin.

10.4 Catalase Test

Some bacteria can reduce diatomic oxygen to hydrogen peroxide or superoxide. Both of these molecules are toxic to bacteria. Some bacteria, however, possess a defense mechanism which can minimize the harm done by the two compounds. These resistant bacteria use two enzymes to catalyze the conversion of hydrogen peroxide and superoxide back into diatomic oxygen and water. One of these enzymes is catalase and its presence can be detected by a simple test. The catalase test involves adding hydrogen peroxide to a culture sample or agar slant. If the bacteria in question produce catalase, they will convert the hydrogen peroxide and oxygen gas will be evolved. The evolution of gas causes bubbles to form and is indicative of a positive test.

10.5 Fluorescence Test

The water soluble fluorescent pigment can be detected using Kings B medium. If the culture produces diffusible water soluble pigment, a pigment zone is produced around the growth, which is easily detected especially when viewed in UV light.

King's B Medium : This is used for growth of *Pseudomonas* and to detect fluorescent pigment formation – Peptone : 20 g, Glycerine : 10ml, K_2HPO_4 : $MgSO_47H_2O$: 1.5 g, Agar : 20 g, Water : 1000 ml.

Table 10.1 Important phytopathogenic bacterial genera.

Characters	Genera				
	Clavibacter	*Erwinia*	*Ralstionial Pszeudomonasi Burkholderia*	*Xanthomonas*	*Agrobacterium*
Disease	Gumming of infloresceneces, Wilts and/or Leaf spots	Vascular wilts, Dry necrosis, Leaf spots, Soft rots	Vascular wilts, Leaf spots, Soft rots	Vascular wilts, Leaf spots, Stem canker	Crown gall, Hairy root formation, Soil saprophyte
Gram reaction	+	-	-	-	-
Motility	Variable, some are motile and some are not	Motile	Motile	Motile	Motile
Flagellation	Polar / lateral (few)	Peritrichous	Polar (one or several)	Polar (one)	Sparse
Percent Guanine + cytosine in DNA	65-75	50-58	58-70	63-69	59-63

Chapter 11

Serological Identification of Plant Pathogenic Bacteria

Serology is the science of reactions, preparations and use of serums. Studies of antigen –antibody reactions constitute the methods of serology. Carpenter defined an antigen as "a substance that elicits a specific immune response when introduced into the tissues of an animal. The response may consist of antibody production, cell mediated immunity, or immunologic tolerance".

Antiserum is the fundamental reagent of serology and provides great versatility and specificity. Antiserum is produced by an animal 's lymphoid system which, recognizing an antigen as a foreign substance and an implied biological threat, manufactures proteins (immunoglobulins) that react selectively with the foreign "body". These proteins carried in body fluids are known collectively "antibodies".

Serology is extremely sensitive. Precipitating antibody can be induced by a single injection of only 5 g of egg albumin. The dose given may be varied according to the purpose for which the antiserum is needed. Optimal production of antibody in rabbits has been show to require only 0.5 mg of a hapten-protein conjugate. If this dose of antigen per injection is used, a 1% contaminant of the inject antigen could induce precipitating antibodies. Since an amount greater than 0.5 mg is normally used for injection, a contamination of the inject antigen of less than 1% would be required to obtain a monospecific antiserum. However, antisera can often be rendered specific for a given antigen by absorption. The unwanted antibodies can be neutralized by adding the respective antigens to the antiserum. For most immunochemical purposes, however, it is preferable to use insolubilized antigens (protein copolymers or proteins conjugated to insoluble matrices) for the absorption. Absorption Techniques are not generally recommended, however.

Antibodies are oblong shaped molecules with active sites at both ends. The antigen–antibody reaction results in the formation of a three dimensional network or lattice in which the antigen and antibody alternate. This is termed a primary antigen-antibody reaction and cannot be observed in usual test tube or agar plate procedures; it can be observed by immunofluorescence. Secondary reactions, called precipitation, require several hours, or even days, depending upon the conditions. Precipitation is final visible result of a serologic reaction. Agglutination is the same as precipitation expcept that the antigen is particulate, e.g., a virus particle or bacterial cell rather than a molecule.

11.1 Serological Tests

The first report about using serology to identify a plant pathogenic bacterium was reported by Jenson (1918). He showed that a strain of *Agrobacterium tumefaciens* from Denmark could be differentiated from a strain of *A. tumefaciens* by agglutination test. Agglutination tests with antiserum (As) that were made by use of live or formalized cell preparations rapidly became very popular during the 1930s among researchers working on the identification of plant pathogenic bacteria and remains today. Such techniques were thought to be specific and were suggested for differentiating or identifying *Erwinia atroseptica, E.aroideae, E.caratovora, E.amylovora, Agrobacterium tumefaciens, Rhizobium* spp., *Cornebacterium sepedonicum, pseudomonas* spp., *P.solanacearum,* yellow pigmented bacteria, *Xanthomonas malvacearum, X .translucens* and a newly described xanthomonad isolated from Filbert.

11.1.1   Agar Precipitation

With the introduction of agar double diffusion precipitation test by Ouchterlony in 1948, agar precipitation tests became widely used during the 1950s for identification of medically important   bacteria. However, Ouchterlony double diffusion (ODD) was not tested for identifying a plant pathogenic bacterium until 1960. ODD has been used to identify or differentiate strains of *P. syringae, P.phaseolicola P. lachrymans, P. savastanoi, P. solanacearum , P.pisi, P.avenae, P.setariae P. glumae,* and *P. andropogonis and pseudomonas spp.* strains of *X. vesicatoria , X.oryzae X. ampelina and X. cyamopsidis.*

Ouchterlony double diffusion tests were used to confirm the identity of a strain of *Rhizobium* isolated from galls as original inoculum strain.

11.1.2   Immunofluorescence

Immunofluorescent (IF) staining was suggested for identification of bacteria in plants in 1943. Morton reported the superiority of direct IF over agglutination or bentonite flocculate tests for rapid identification of *X. vesicatoria* in extracts of diseased tissue. In 1967 indirect It was used to to identify *E. aroideae* in extracts of host tissues and soil. It has been used to identify *P. phaseolina* in preparations of bean seeds and leaves, *P. solanacearum* in soil, *P. tabaci* in leaves and agar culture, *E. atroseptica* in culture and in preparations tubers, leaves, insects and soil, *X. albilineans* in host tissues, *X campestris* in culture and preparations of soil, *C. michiganense* in seeds, *Spiroplasma citri* in host tissues.

Serology is an extremely valuable tool in molecular biology, Serological techniques are presently used for purification, determination of structure and function and identification of specific compounds. Serological assays were originally developed to detect viruses, since they could not be cultured. Formats for immunodiagnostic techniques include enzyme-linked immunosorbent assay (ELISA), immunofluorescence (IF), immunflourescence colony staining, and immuno-strip tests. ELISA, by far the most common immunodiagnostic technique, has been consistently used for virus and bacteria detection since the 1970s, long before DNA- based techniques were available. There are over 800 different antisera available for plant viruses through the American Type Culture Collection polyclonal and monoclonal antisera for many viruses and bacteria have been developed for commercial use or in individual labs. These antibodies have been used in numerous protocols to identify viruses, including immunodiffusion assays, western blots, dot-blot immunobinding assays, immuno-strip assays, and serologically specific electron microscopy (SSEM). However, ELISA remains the consistent protocol of choice for viruses and bacteria in diagnostic labs due to its high throughput capability.

The essential of an ELISA assay varies depending on the organism, Commercial ELISA kits are available for many viruses, bacteria, and fungi (Agdia, Elhart, IN and ADGEN, Ayr, Scotland). The typical sensitivity of ELISA for bacteria is 10^5 CFU/ml. The technique works best for diagnosis when samples consist of fresh lesions containing very high pathogen titers.) IF techniques use fluorescent-labeled antibodies that react either directly with cell antigens or indirectly with anti-rabbit globulin and allow for direct visualization of cells with a fluorescent microscope. The technique has an advantage in that cell morphology is observed. IF is especially useful for detecting seed borne and tuber-borne bacteria, but is reliant on human judgement to determine whether a cell is fluorescent. IF can also be used to identify colonies of bacteria in agar media, which can be advantageous since it is based on viable cells. For rapid presumptive identification, immuno-strips can be very useful. The technique is inexpensive and requires little labor or knowledge. Specificity of all immunoassays can be improved by using monoclonal antibodies. However, increased specificity means that some target strains may be missed (false negative). Bacteria, which are very complex and heterogeneous with respect to surface antigens, can vary with the environment in which the cells are growing. Serological techniques can greatly reduce the time needed for diagnosis; however, the results should only be considered as presumptive since both false positives and negatives are possible.

Chapter 12

PCR Based Diagnosis of Bacterial Diseases

PCR has revolutionized molecular biology and diagnostics. A rapid serological dip-stick technique followed by PCR has been the preferred method for detection of spores of anthrax. With the development of PCR, DNA-based techniques have rapidly become the preferred tool for identification of plant pathogenic bacteria. A large number of classical PCR primers for identification of important plant pathogenic bacteria are listed in the Laboratory Guide for Identification of Plant Pathogenic Bacteria and many more are becoming available, PCR-based assays offer many advantages over traditional isolation and immunological methods; most important are specificity and time.

Specificity of PCR depends upon the uniqueness of the sequences selected for primers and probes. Improvements in sequencing technologies are making the selection of reliable PCR primers routine.

(i) *For PCR based Diganosis of Bacterial Pathogens (Ralstonia solanacearum)*

    (a) Prepare the PCR reaction mixture (optimized for the sample), separately in a thin wall 0.5µl micro centrifuge tube in the order listed below:

Constituents	Per 25 ul reaction Vol.
Deionised nuclease free water	13.5 ul
10X PCR Buffer	2.5 ul
DNTP	2.0ul
$MgCl_2$ (25mM)	2.5ul
Primer F (RS Y1)	0.5ul
Primer R (RS Y2)	0.5ul
Taq DNA Polymerase (5U/ul)	0.5ul
DNA sample Template (25-50ng)	3.5ul
Total Volume	25.0ul

Mix gently and carry out the polymerase chain reaction in the thermal cycler with the following specifications:

Temperature (Deg C)	Time (secs)	Cycles	Purpose/Function
94° C	3 min .0 sec	1	Initial Denaturation
94° C	60 sec	35	Denaturation of template
60° C	60 sec		Primer annealing
72° C	80 sec		Extension
72 °C	8 min	1	Final Extension

12.1 Real-time Fluorescent-based PCR

Real time PCR can use TaqMan ^{Tm} probes, fluorescent resonance energy transfer (FRET) probes, or molecular beacons to detect the production of amplicons. These methods are based upon the hybridization of fluorescently labeled oligonucleotide probe sequences to a specific region within the target amplicon that is amplified using traditional forward and reverse PCR primers. In the TaqMan ^{TM} system, an oligonucleotide probe sequence of approximately 25-30 nucleotides is labeled at the 5' end with a fluorochrome, usually 6-carboxyfluorescein (6-FAM), and a quencher fluorochrome, usually 6-carbosy-tetramethyl-rhodamine (TAMRA), at the 3' end. The TaqMan ^{TM} probe is degraded by the 5' to 3' exonuclease activity of the Taq polymerase as it extends the primer during each PCR amplification cycle and the fluorecent chromophore is released. The amount of fluorescence is monitored during each amplification cycle and is proportional to the amount of PCR product generated.

FRET probes require labeling of two adjacent oligonucleotide probe sequences within the PCR target fragment. Probe 1 contains a fluorescein label at its 3' end, whereas the second probe is labeled at its 5' end with another label such as Light Cycler Red 640. The two probes must be designed so that when they hybridize to the amplified PCR product they are aligned head-to-tail to bring the two fluorescent dyes in close proximity to each other. The fluorescein dye attached to the first probe is excited by the light source of the appropriate wavelength, and it emits a green fluorescent light at a slightly longer wavelength. When the second probe is in close proximity, the energy emitted by the first probe excites the Light Cycler Red 640 dye attached to the second probe, and red fluorescent light at longer wavelength is now emitted to that can be detected at 640 nm. Fluorescence is measured during the annealing step of each of the amplification cycles when both probes hybridize to the PCR amplicon (16,20). Molecular beacons are hairpin-shaped fluorescent oligonucleotide probes. The loop portion of the molecule contains nucleotide sequences that are complementary to the target amplicon. A fluorescent chromophore is attached at the 5' end of the probe and a quencher molecule is attached at the 3' end. A stem structure is formed by annealing of the complementary arm sequences that are added on both sides of the probe sequence. When a stem structure is formed, the fluorophore transfer energy to the quencher. However, when the probe hybridizes to the target amplicon during PCR amplication, the fluorophore and quencher become separated from each other and fluorescence can be detected.

The first platform designed for real-time PCR, AB17700 Sequence Detection System [R] from Applied Biosystems (Foster City, CA), allows upto 96 sampes to be run simultaneously and provide endpoint data analysis within 2 to 3 h. Its high throughput capacity makes the AB17700 well suited for use in the pharmaceutical industry, but its very high cost (over \$US 80,000) is a deterrent to routine diagnosis of plant diseases. By using real-time TaqMan PCR and Perkin Elmer 7700 Sequence Detection System (Applied Biosystems, Foster City, CA), the time for PCR can be reduced from 2 days to 2-3 days to 2-3 hours. Due to its very high throughput, the 7700 is very cost-effective when used routinely for large numbers of samples.

Perhaps the most significant improvement in real-time PCR for diagnostics has been the development of relatively inexpensive, ultra-fast portable thermocyclers. Several such instruments are available for performing rapid-cycle real-time PCR, including the Light Cycler [TM] from Roche Diagnostics Corporation (Indianapolis, IN). The R.A.P.I.D. from Idaho Technologies (Salt Lake City, UT), the SmartCycler[R] from Cepheid, Inc. (Sunnyvale, CA), the iCycler iQ[TM] from Bio-Rad (Hercules, CA), the MX4000[TM] from Strategene (La Jolla, CA), the Rotor Gene, from Corbett Research (Sydney, Australia), and the Gene Amp 5700[R] from Applied Biosystems (Foster City, CA). Except for the Gene Amp 5700[R] instrument, these units are capable of data collection (monitoring fluorescence) during each cycle of PCR amplification. Real-time PCR has many important advantages over classical PCR: it yields data in real-time, is quantitative, and does not require a separate step to detect amplified products (agarose gel electrophoresis); results are based on hybridization of the probe to the target amplicon, i.e., a time-consuming Southern blot is not needed for maximum sensitivity and for confirming the identify of the amplified product.

12.2 BIO-PCR

Classical and real-time PCR protocols are available for many different bacteria. Although real-time PCR is fast and can be very specific, the technique is often not as sensitive as agar media-based assays to detect pathogens present in plant extracts, as inhibitors and small sample volumes can reduce sensitivity. However, PCR can be used in combination with isolation on media, a technique termed BIO-PCR whereby viable cells of the target bacterium can be enriched in liquid or solid media and detected in extremely low levels in seeds and other propagative materials. In a BIO-PCR assay, the plant extract is plated onto agar or added to liquid media and enriched for 15 to 72 h, depending upon the organism, and the resulting cell growth used for direct PCR. No DNA extraction is needed for bacteria since the cells will lyse during the initial denaturing step of the amplification. BIO-PCR protocols have been developed for several bacteria including *P.syringae pv. phaseolicola* (72,74), *Clavibacter michigenesis subsp. sepedonicus, Acidovorax avenae subsp. avenae, A.tumefaciens,* and *E.coli*. When time is more important than sensitivity, BIO-PCR is not recommended. For ultra-high sensitivity, BIO-PCR can be used with membranes. Using surfactant – free membranes (Sartorius, Edgewood, NY, no.12587) made especially for BIO-PCR, the sample is filtered to retain the target organism and the membrane is placed on semiselective agar media. After incubation for 1 to 3 days, the filter is removed and placed into a microfuge tube with 50-ul water and vortexed to suspend the bacteria. The sample is then used for direct PCR. This works well for samples such as leaf washings or pond water. With P.syringae pv. phaselicola, 1 to 3 cfu/ml water are routinely detected. Membranes can also be used for standard PCR without waiting for BIO-PCR. An added advantage of membrane PCR is that the sample is freed from PCR inhibitors. Membranes can also be used successfully with direct PCR (without enrichment) by placing membranes of smaller diameter directly into the reaction tube. The main disadvantage of using membranes in PCR is that the technique introduces a step for possible cross-contamination. An alternative to BIO-PCR for increasing the sensitivity of PCR is to use immunocapture PCR.

Real-time BIO-PCR protocols have detected *P.syringae* pv. *phaseolicola* in seed extracts and *C.sepedonicus* in potato tuber extract at a concentration of 2 cfu/ml. Of 30 naturally infected potato tubers assayed by agar plating and conventional real-time PCR, 4 and 8 were positive, respectively. In contrast, with BIO-PCR 26 of the 30 tubers were positive. Similarly, Weller et al. found that conventional real-time PCR worked well. In addition, multiplex PCR can be employed to detect more than one species of bacterium in the same reaction tube using probes labeled with different fluorescent dyes such as FA, TET, TAM, and ROX. To avoid cross-absorption, the wavelength of each dye must be well separated.

A number of techniques, including hundreds of electron microscopy, nucleic acid hybridization, ELISA, and PCR-based protocols have been used to detect viruses in crops. The most commonly used survey and indexing protocols now take advantage of ELISA or PCR techniques. ELISA detection kits are readily available for a large number of viruses, and are often favored by diagnostic labs for their speed and relative cheapness, despite their lack of sensitivity. Many viruses exist at low or variable titer levels that are hard to consistently detect using ELISA. In addition, the time involved in good antibody production, the possibilities for false positives, and the inability to differentiate between closely related viruses also detract from the effectiveness of ELISA detection protocols. Finally, ELISA cannot detect some of the important pathogens that exist solely as nucleic acids, such as Viroids.

Table 12.1 Amplification of bacterial target sequences by PCR.

Organism	Host/Disease	Target sequence	Comments/References
Agrobacterium tumefaciens	Tumors (galls) Vitis spp. (grape)	IaaH and iaaM pathogenicity genes of Ti plasmid; DNA extracted from purified isolates	With most isolates the presence of Ti pathogenicity genes (positive PCR results) correlated with pathogenicity tests
Erwinia amylovora	Pome/fire blight	Plasmid sequences ; boiled bacteria and infected plant tissue; inclusion of Tween 20 in reaction buffer increases sensitivity.	Tth polymerase used
Mycoplasma like organisms (MLOs)	Various	Conserved rDNA sequence; DNA extracted from infected plants	Detects many MLOs; use RFLP to distinguish among isolates
Mycoplasmalike organisms (MLOs)	Periwinkle	Cloned MLO sequences; DNA extracted from infected plants	Detection of MLO DNA
Pseudomonas solanacearum	Bacterial wilt/wide host range	Specific sequences cloned by subtractive hybridization; boiled bacteria and infected plants	Detection of 5-116 bacterial cells, depending on the strain, also sensitive detection in potato field trails
Rhizobium leguminosarum	Nitrogen fixation/beans	npt gene of Tn5; boiled nodule tissue; DNA extraction from soil	Detection in nodules and soil; other bacteria harboring Tn5 would also be detected.
Xanthomonas campestris pv. citri	Citrus/bacterial canker	Plasmid sequences with high G+C; content; either purified bacterial DNA or boiled, crude water extracts in buffer containing Triton-100	Target sequences detected in lesions, from which viable bacteria could not be cultured

12.3 Three D's of PCR–based Genome Analysis for Bacteria

Three Ds-diversity, detection, and diagnosis – provide a framework for the application of different PCR-based genome analysis protocols. "Diversity" refers to the degree of genetic variation within bacterial populations and relates to bacterial systematics, at multiple taxonomic or phylogenetic levels, as well as to the structure of pathogen populations. Our key premise is that assessing genetic diversity of populations is required to established a stable taxonomy and developing durable disease management strategies for pathogenic bacteria. Such an approach is analogous to mapping soil variation that provides global position coordinated for precision agriculture;

"Detection" deals with establishing the presence of a particular target organism within a sample, while "diagnosis" relates to the "identification of the nature and cause of a disease problem". Identification typically involves assigning an unknown organism to a known taxonomic group, based on selected characteristics (158). Regardless of the characteristics chosen, plant disease diagnosis and pathogen detection approaches require substantial levels of specificity, sensitivity, and speed. Specificity, in this case, is defined as the capability to target the organism of interest, in the absence of false positives when it is not present or false negatives when it is present. In the case of PCR-based methods, this specificity is largely determined by primer selection and amplification conditions. Sensitivity relates to the lowest number of pathogen cells per assay or sample that can be reliably detected and depends on the nature of the PCR protocol, including sampling protocols and sample preparation. Plant disease diagnosis is more dependent on specificity than on sensitivity. Usually, diagnosis is based on observed symptom and generally only a limited number of known possible casual agents need to be considered. The titer of the pathogen is generally higher in symptomatic tissues than in seed lots, for example, and in many cases, the pathogen can first be cultured on selective media and only suspect colonies are subjected to PCR-based identification protocols. Identification of a pure culture can be accomplished by using specific primers to amplify a diagnostic fragment or universal primers that provide a diagnostic genomic fingerprint.

PCR-based diversity studies are generally accomplished with universal primers that generate an array of DNA amplicons frequently referred to as genomic fingerprints. Three protocols have been commonly employed to fingerprint phytobacteria, including arbitrarily primed [AP-PCR] or randomly amplified polymorphic DNA [RAPDs], respective sequence-based rep-PCR, and amplified fragment length polymorphism [AFLP] analysis. Each of these protocols generates a collection of genomic fragments via PCR, which are resolved as banding patterns that provide a high level of taxonomic resolution.

12.4 AP-PCR and RAPD-Based Genomic Fingerprinting

AP-PCR/RAPD analyses have been used extensively to assess the diversity or nature of phytobacteria. Short oligonucleotides of an arbitrary DNA sequence, used in pairs or singly under low stringency conditions fortuitously anneal to target DNA sites that are partially or fully complementary to the primer sequence. An array of variable-size PCR products are generated corresponding to the number of annealing site pairs proximal enough for PCR-mediated DNA amplification. AP-PCR/RAPDs requires no prior knowledge of target DNA sequences but may require empirical screening of a number of arbitrary primers to generate the genomic fingerprints desired.

12.5 Repetitive DNA PCR-Based Genomic Fingerprinting

Rep-PCR analysis was developed based on the observed occurrence of specific conserved repetitive sequences [repetitive extragenic palindromic (REP) sequences, enterobacterial repetitive intergenic consensus (ERIC) sequences, and BOX elements] distributed in the genomes of diverse bacteria. However, the term has been expanded to include the use of primers for PCR genomic fingerprinting that anneal to any three primer sets are commonly used for rep-PCR genomic fingerprinting analysis, corresponding to REP, ERIC, and BOX sequences. A complex array of 10 to 30 or more PCR fragment is generated per genome, ranging in size from less than 200 bp to more than 6 kp Rep-PCR has been extensively used to identify pathogens, to differentiate strains, and to assess the genetic diversity of plant pathogens each primer set (REP, ERIC and BOX) is useful to fingerprint diverse bacteria, including gram-negative and gram-positive phytobacteria, as well as plant-associated actinomycetes.

12.6 Amplified Fragment Length Polymorphism Genomic Analysis

AFLP analysis also appears to be a universal protocol to fingerprint plant pathogenic bacteria restriction of genomic DNA using two restriction endonucleases, ligation of specific double-stranded DNA adapters to the respective restriction fragments to function as priming sites, amplification of fragments using two primers complementary to the ligated adapters. The primers include one or two additional nucleotides at the 3' end designed to selectively match genomic DNA sequences flanked by the adapters in order to generate specific fragment sets that perfectly match the adapters and adjacent bacterial nucleotide(s).

Impacts of Molecular diagnostics in plant disease management

12.7 New, Emerging, and Introduced Pathogens

The greatest threats to agriculture have always come from nature: freezing temperatures, hail, high winds, flooding, droughts, and pests and pathogens. New and emerging diseases are an ever-increasing reality for phytopathologists. To bring along live plants or plant materials that may harbor exotic pests and pathogens.

The deliberate release of a crop pathogen is not a new threat. Deliberate releases of plant pathogens onto crops have been reported since biblical times. The Former Soviet Union (FSU), the United States, Canada, Germany, Japan, China, South Africa, and Britain are known to have had major crop biowarfare programs.

Despite the historical record of biological agents as weapons the potential scope of the deliberate release of a crop pest or pathogen has been realized only relatively recently. The likelihood of major deliberate release in the United States was deemed remote before the anthrax releases in October 2001 dramatically brought home the reality and need for increased bio security.

An accidental or intentional introduction of a regulated plant pathogen could easily cause losses to the economy in tens of billions of dollars.

The protection of animals and crops from accidental or deliberate introduction of pathogens is an important national security priority for any country indeed. Recent technological advances that allow for the genetic manipulation of pathogenic virulence and other characters have significantly raised the threat of a deliberately released foreign pathogen to become established. The number of high-risk pathogens identified today outstrips those that were known during world war II and the cold war.

The food production and distribution network is susceptible to contamination with human pathogens such as *Escherichia coli* and *Salmonella* species. *E.coli* 0517:H7 increases in numbers very rapidly on small fruits and vegetables following infection with plant pathogens. If no plant pathogens are present, numbers of *E. coli* remain below thresholds needed for human infection. Also, plant pathogens themselves can pose real threats. Several plant pathogenic fungi and bacteria produce animal and human toxins in plants. Hybrid vegetable seed which is mostly produced abroad by local contract growers, can serve a major avenue for accidental or deliberate introduction of seed borne pathogens. An important fresh market vegetable threatened by a devastating seed borne bacterial disease-watermelon fruit blotch caused by *Acidovorax avenae subsp. citrulli.*

How do we protect out crops and crop products against introductions that may manifest as subtle, even latent, disease outbreaks? The phytopathologist has much to contribute, in well through collaboration with other professionals, to the strategic protection of crops against the accidental or deliberate introduction of foreign pathogens. Soft targets such as agriculture do have consequences, especially when one considers the economic impact in potential losses in trade.

12.7.1   Identifying High Risk Pathogens

One of the best defenses against the introduction of new plant diseases, by either accidental or deliberate means, is rapid detection. One mission of the plant pathology research program of the USDA/ARS Foreign Disease-Weed Science Research Unit (FDWSRU) at Ft. Detrick is development of rapid molecular based systems for detection of naturally introduced foreign pathogens for use by Animal Plant Health Inspection Service (APHIS), DOD, other federal and state agencies, and universities. There are over 500 pathogens that can cause major disease losses; a reliable methodology for rating and prioritizing those pathogens of the highest risk is therefore essential. Madden et al, have discussed the importance of Epidemiological data in establishing lists of high-threat pathogens.

12.7.2   Diagnosing Diseases Under Field Conditions

Presumptive diagnosis of plant diseases in plants showing symptoms can be relatively simple when typical, definitive symptoms are evident. However, symptoms are not always unique and can be confused with other diseases. Typical examples are halo blight of beans, caused by the regulated *Pseudomonas syringae pv.phaseolicola.* and brown spot of beans, caused by unregulated organism *P.syringae pv.syringae.* The lesions can normally be differentiated on green leaves because of the yellow halo produced by *P.syringae pv. phasaeolicola.* However, no halos are visible on dried pods; both pathogens produce indistinguishable brown spots. Diagnosis of plant diseases can be even more difficult with infected seeds asymptomatic infected propagative materials such as tree grafting stocks or potato tubers. Traditional isolation and pathogenicity tests require 10 to 20 days or longer, enough time for bacteria and aerial fungi to spread dramatically, causing severe epidemics.

Chapter 13

Transmission of Viruses

Bio- assay of plant viruses is essential to determine the presence of virus in the plant extract ,amount of virus present ,whether virus multiplies in the vector or not, rate of multiplication and its persistence in the host plant. There are several methods used for assaying ,include immuno-electron microscopy (IEM),Enzyme linked immunosorbent assay (ELISA), Leaf dip serology etc. Among these, bio-assay method has the advantage of indicating whether the virus is infective or not.

Transmission of viruses can be grouped into :

  (a) Mechanical transmission

  (b) Transmission through insects

  (c) Grafting

Local lesion hosts are those which react to mechanical inoculation with several viruses by producing countable infection sites called local lesions.

Differential host that can distinguish between certain viruses when present together in the same host plant

Indicator plant host are those reacts diagnostically with certain viruses when present together in the same host plant.

Based on the level of resistance to a particular virus, plants respond differently to different viruses. Local lesion has got greater significance for assessing the presence of plant viruses.

The susceptible plants react to virus either by producing local lesions, in which the virus infecting is confined to the initial infected tissues or by producing systemic symptoms where the symptoms appear away from the site of infection or inoculation.

13.1 Mechanical Transmission

Many viruses can be transmitted mechanically from suspected host plant to indicator or test plants. Such transmission of viruses requires successful extraction of the virus from the host material and transfer of virus bearing sap to the surface of the leaves of test plant in such a way that viruses can enter the cells and produce symptoms .Local lesions are produced ,if the inoculated plant is susceptible.

The infected leaf is grinded and the crude sap is diluted serially and inoculated on to the leaf of a test plant mechanically using an abrasive like carborundum. The number of local lesions produced by a particular dilution is a measure of the amount of virus present. The lesions will gradually appear in about two days after inoculation and in some cases it will take 6-7 days. The local lesion hosts are used for both detection and assay of plant viruses.

Materials Required : Test plant (*Nicotiana glutinosa* or *Chenopodium amaranticolor*), carborundum powder, Pestle and mortar ,Phosphate buffer 0.05M(pH 7), Cheese cloth, Scissors etc.

Procedure : Select the test plant (2 no.) of each species and two oldest, fully expanded healthy green leaves are chosen for inoculation. Take the infected leaf material into a mortar and add 1ml of phosphate buffer 0.05 M (pH 7) per gram of leaf and grind it till a fine homogenate is obtained , then filter the extract through the muslin cloth. Dust the carborundum powder (600 mesh) finely on test plant and rub gently the inoculum on the leaves of test plant with the help of cotton swab. Keep the plants in glass house and observe the local lesions at periodical intervals   and record the results.

13.2 Insect Transmission

Various groups of insects such as white flies, aphids, thrips, mites etc transmit plant viruses and viruses being transmitted by a particular group have the difference in number of other properties. The nature of vector involved in transmission and the relationship between the virus and vector has been studied for majority of plant viruses and MLOs

13.2.1   Aphid Transmission

Based on the difference in properties of viruses in vectors like persistence, latent period, multiplication of viruses in aphids , the aphid borne viruses can be grouped into non-persistent, semi persistent, propagative and non-propagative viruses.

Table 13.1. Differences between groups of viruses

Character	Non-persistent	Persistent	Propagative	Circulative
Acquisition feeding period	Short	Long	Longer latent period	Defined latent period
Efficiency	Shorter period of starvation increase acquisition	No influence of fasting	-	-
Incubation period	Shorter	Long	Multiply in vector	Donot multiply in vector
Retention	Shorter period	Longer period	-	-
Examples	Potato virus S,M(Rigid rods) Potato virusY, Beet mosaic (long flexible rods) Cucumber mosaic (Isometric)	Citrus tristiza virus ,Beet yellows	Lettuce necrotic yellows	Pea enation mosaic virus, Luteo virus

Aphids form an important group among insect vectors. Among insects, aphids have evolved to be most successfully exploiting higher plants as food source, particularly flora of temperate regions.

Materials required : Test plant (virus infected and healthy), Aphids (must be non viruliferous) parafilm, camel hair brush ,filter paper etc.

Procedure : Carefully pickup wingless aphids and transfer them to Petridish lined with moist filter paper with the help of camel hair brush, then close Petri plate with parafilm. Keep the plates in a cool shaded area for about 1hr to starve them. Then open the plate removing parafilm and transfer aphids to infected plant material and allow to feed for different time intervals. Care should be taken to transfer atleast five aphids to each test plant. A polythene cover is used to contain aphids and allow a transmission feed of not more than 1 hour. After that, kill vector aphids by using insecticide and place the test plant in glass house to grow-on for symptom observation.

13.2.2   Leafhopper Transmission

The leafhopper borne viruses can be divided into three groups namely semi-persistent, circulative and propagative viruses based on transmission character.

Table 13.2 Examples of viral diseases transmitted by Leafhoppers.

1.	Rice Tungro virus	*Nephotettix virescens*	Semi-persistent
2.	Maize chlorotic dwarf virus	*Graminella nigrifrons*	Semi-prsistent
3.	Potato yellow dwarf virus	*Agallia constricta*	Propagative
4.	Maize rayada fino virus	*Dalbulus maidis*	Propagative
5.	Rice stripe virus	*Laodelphax striatellus*	Propagative

Materials required : Insect proof cages, aspirator, forceps, seed of a rice variety susceptible to tungro (Ex:TN1), Rice plants suspected to be infected by tungro virus, test tubes with plastic caps.

Procedure : To demonstrate leaf hopper transmission, the rice variety TN1 is chosen because of its susceptibility to Tungro virus. Pullout the rice seedling susceptible to RTV (TN1) and place it in a test tube containing west sand. Collect non-viruliferous leaf hoppers and subject to 2 hrs. of starvation and then allow it to feed on suspected virus infected plant for 24 hrs. Two viruliferous green leaf hoppers will be inoculated per seedling for 24 hrs and cover with muslin cloth and plant the seedlings in pots, observations will be made for symptom development. If the test plants are being infected by Tungro virus, symptoms will be developed within 12 to 15 days.

13.2.3   Whitefly Transmission

Whiteflies are known to transmit about 70 viruses to mainly tropical and subtropical plants. Many of the viruses transmitted by whiteflies cause mosaic disease of a bright yellow or golden nature. Less commonly, the diseases involve marked curling of leaves or generalized yellowing. Many of the Gemini viruses and some viruses belonging to carla virus, clostero virus, nepo virus and poty virus group are transmitted by whiteflies.

Only three species of whiteflies have been reported as vectors of plant viruses and *Bemisia tabaci* is the vector for many diseases. None of the viruses transmitted by whiteflies has non-persistent relationship with its vector and most of the viruses are of circulative (non-propagative) and no multiplication in the vector.

Materials required : Mylar cages, whiteflies (*Bemisia tabaci,*) healthy mungbean / tomato plants; yellow mosaic infected mung bean / tomato leaf curl infected tomato plants.

Procedure : Collect the non-viruliferous whiteflies with the help of an aspirator into test tubes grown under glasshouse conditions and cover the tubes with muslin cloth. Let them starve for 2 hrs under shade in cool conditions. After starvation, release the whiteflies onto the yellow mosaic infected plants or tomato leaf curl infected plants kept in mylar cages. Allow the whiteflies to acquire virus for 24 hrs. After 24 hrs, with the help of aspirator, collect the viruliferous whiteflies and transfer to healthy tomato or mung bean plants at the rate of three whiteflies per plant. The plants are to be maintained in an insect proof cage. Allow the viruliferous whiteflies to feed on healthy plants for 12 hrs and after that they are killed using insecticide. The plants are to be observed for the development of symptoms which usually appear after 10 to 15 days.

13.3 Grafting

One of the important characters of a plant virus is its ability to transfer from infected to healthy plant when both are grafted together. For many years such transfer was the only means of demonstrating the presence of a transmissible agent, for many virus-host combinations, this still remains a valuable test. Viruses which become systemic in their host can usually graft transmitted provided the donor and recipient host are compatible.

Grafting plays an important role in the identification of plants infected either by virus or by MLO's but not by fungi or by bacteria. In this method all the viruses which are systemically distributed can be transmitted to susceptible host plant by grafting. In this method, the scion and stock of the graft get organically connected to each other to form continuous channel of xylem, phloem and plasmodesmata for the transmission of nutrients, and at the same time viruses are also carried over. Plasmodesmata are formed soon after the union of parenchymatous cells while vascular tissues are connected few days later. Parenchyma restricted viruses pass over earlier than the phloem restricted viruses. Grafting plays an important role in propagation of fruit and ornamental plants but some times the viruses infecting the crop are also transmitted. The notable example is the citrus tristeza virus disease of sweet oranges.

Materials required : Virus infected plant (Scion), healthy susceptible plant (Root stock), Blade or sharp scalpel, polythene strips, Glass tubes 20 × 60 mm.

13.3.1   Cleft Grafting

Select the donor plants which are actively growing and choose recipient or 'scion' shoot of approximately similar size. Top portion of root stock plant is removed by making two downward oblique cuts. Similarly, top part of the scion or indicator plant is removed making downward cut at a point where the stem diameter is similar to that of the stock. Insert scion shoot into stock stem so that they fit well and there after, the union is firmly bound with polythene strips. If necessary, support the scion with a stick and observe for development of symptoms.

Serological Identification of Plant Viruses

Serology is an extremely valuable tool in detection of plant pathogens. Serological techniques are presently used for purification, determination of structure and function and identification of specific compounds. Serological assays were originally developed to detect viruses, since they could not be cultured. ELISA, by far the most common immunodiagnostic technique, has been consistently used for virus and bacteria detection since 1970's long before DNA based techniques were available. There are over 800 different antisera available for plant viruses through the American type culture collection. Polyclonal and monoclonal antisera for many viruses have been developed for commercial use. These anti bodies have been used in numerous protocols to identify viruses, including immuno diffusion assays, western blots, dot-blot immunobinding assays, Immuno – strip assay and serologically specific electron microscopy (SSEM). However, ELISA remains the consistent protocol of choice for viruses and bacteria in diagnostic labs due to its high throughout capability.

For detection of plant viruses, application of serology is the most quick and reliable method which are based on the detection of specific antibodies. The specific relation between antibodies and their homologous antigens was recognized in 1950's. Various diagnostic tests often provide valuable clues to determine etiology, but every test In vitro has its intimate limitations. Based on nature of pathogens, the applicability of serological tests will vary. Some are better than other, in the sense that results correlate better with the diseases. Some are reliable and have proven merit. Success in execution and interpretation of sero-diagnostic tests are mainly dependent to a large extent on the experience gained by the individual or group with their systems and reagents. The pre-requisite for performing any serological reaction or test is the production of antiserum. The procedure for production of antiserum is briefly described in chapter 14.

Chapter 14

Production of
Polyclonal Antiserum

The preliminary requirement for performing any serological assay is production of antiserum, either monoclonal or polyclonal. Antiserum is obtained by injecting a protein or virus into an animal over a period of several months and then harvesting some blood from the animal. Antibodies or immunoglobulins, a special kind of proteins are produced in response to the presence of foreign molecules. These proteins bind specifically to the molecule that induce their formation, referred to as antigen and an animal serum containing antibodies is referred to as an antiserum. When the blood is collected from animal, allowed to clot from which serum is separated. This serum fraction contains antibodies, which can be used directly in immunoassays or processed to purify the immunoglobulins. Many substances are known to stimulate the production of antibodies either through injection or infection, however the prerequisite being the molecular weight structure (10^4 Daltons). There are mainly five types of classes (Ig G, IgM, IgA, IgE and IgD) and eight subclasses of antibody molecules. Of these, IgG is the main serum antibody The antigen – antibody interaction involves hydrogen bonds, electrostatic charges, salt bridges and Vander walls forces. The reaction is reversible and highly influenced by physical factors such as temperature, pH and solvent conditions.

14.1 Structure of Antibody

Among different classes of antibodies, Ig G is the largest fraction of immunoglobulins and it may make as much as 75 percent of the total immunoglobulins in serum. The IgG antibody molecule consists of two light chains and two heavy chains held together in a 'Y' formation by disulfide bonds. IgG is a bivalent molecule having an antigen combining region (Fab) on each variable region of each leg of the molecule. The Fc region of the antibody molecule is constant for each antibody class. It is this portion of the antibody to which the enzymes are coupled in enzyme labeled immunoassays while the fab portion retains its ability to bind the antigen.

14.2 Antigen Preparation

For antibody production, it is essential to use highly purified virus preparation to avoid untargeted antibody production against host impurities Antigen are generally of high molecular weight and are commonly proteins or polysaccharides in nature. Polypeptides, lipids, nucleic acids and many other materials also function as antigens. Purified viral inclusion bodies and viral protein such as replicase, movement protein, helper component etc can also be used as antigens. Since, viruses have nucleocapsids as proteins, the purified samples of viruses can be used as antigen.

14.3 Immunization of Animals

For the purpose of antibody production different animals such as rabbits, guinea pigs, mice, chickens and horses may be used but among them, rabbits are the most commonly used because of their easy handling and yield adequate volume of high titred antiserum. Newzealand white rabbits are preferable for raising polyclonal antiserum against viral antigens.

The antigen is administered into animal using different methods like intravenous, intramuscular, intradermal, intraperitonial etc. In intravenous method, antigen is administered with 0.85% saline while in other methods, antigen is mixed with adjuvant in 1:1 ratio and mixed thoroughly to get a clear emulsion before injecting into the animal. The intravenous injection results in the rapid distribution of antigen but the antigen may be degraded quickly. The titre value of antibodies dependent in number of injected. Usually, 2-3 intramuscular followed by one intravenous injection @ 1-2 mg each given at weekly intervals.

14.4 Use of Adjuvants

The most commonly used adjuvant is Freund's complete adjuvant into which soluble antigen is combined as a stable water-in-oil emulsion. The mode of adjuvant action as that it reacts with the antigen in a slow manner and induction of antibody production by *Mycobacterium tuberculosis* present within the complete adjuvant. Freund's complete adjuvant is a mixture of oil and detergent containing *Mycobacterium tuberculosis* where as incomplete adjuvant is mixture of oil and detergent alone.

Materials required : Freud's complete and incomplete adjuvants white Rabbits, Glycerol, Ethanol, Sodium azide, Razor, Butterfly needle and syringe.

Procedure : The purified virus sample (500µl) emulsified with Freund's complete adjuvant (1:1 v/v) is injected intramuscularly into the thigh muscle of white rabbit. After one week, using Freund's incomplete adjuvant inject the emulsion into opposite flank of previously used. Similarly, five to six injections are to be administered. Two weeks after last injection, with the help of fine blade or razor shave the fur from the region over the central ear artery or marginal ear vein. To stimulate supply of blood, apply a drop of ethanol over shaved area. Insert a 21-gauge butterfly needle and allow the blood to flow freely and collect in a glass container. Allow the blood to clot at room temperature in slanting position for 3-4 hr. and then at 1°c overnight in the same position. Collect the supernatant serum in appendorf tubes, having yellow colour. To get rid off any blood cells present in serum, centrifuge at 5000 rpm for 10-15 min. Repeat the procedure four or five times at weekly intervals.

Since the antibodies are protein molecules, get easily denatured. Using appropriate storage methods, such as storing at low temperature (4°C) using antibacterial preservation such as sodium azide 0.02% the serum can be preserved.

Chapter 15
Immunological Techniques for Identification of Viruses

Each virus contains its own specific antigens, differ from other viruses also the host in which it grows. An antiserum designed for one virus would not react with other viruses. Under *in vitro* conditions, a number of tests involve use of agar or agarose gels through which antigen and antibody diffuse and precipitation bands are formed where the constituents meet at suitable concentrations. Such tests separate mixtures of antigen and antibodies by their size, concentration and diffusion coefficient. Thus, these tests are extremely useful in virus identification from crude sap or purified samples.

Significance : Serological tests are helpful in quantitative estimation of virus concentrations and their purification. The distribution of viral antigens and plant tissues in order to locate the presence of virus within a cell can be studied. It forms an important diagnostic tool in identification and differentiation of virus strains.

15.1 Agglutination Test

In this techniques, there will be clumping of antigen when antigen suspension is added to its antiserum and allowed to incubate for some time. The clump or agglutinin that has formed would settle down at the bottom of test tube. This agglutination reaction is favoured by ions of saline reaction in which agglutination takes place.

Chloroplast

This technique is employed in case of elongated viruses where it occur in high concentrations such as potato virus x and Tobacco mosaic virus. A drop of fresh sap from suspected plant is mixed with a drop of antiserum. If antigen and antiserum found homologous with each other, agglutination of chloroplasts is seen. This is a very simple test to perform.

Latex Agglutination Test

In this test, antigen or antiserum is allowed to adsorb on latex polysterene particles when mixed with its antiserum / antigen. These sensitized particles would agglutinate or form a floccule, becomes visible in the form of granular white precipitate within few minutes to one hour. This test has been most commonly used to detect presence of viruses in infected rice plants such as rice tungro, rice grassy stunt etc.

15.2 Precipitation Tests

These are the most traditional methods have been used for direct observation of specific precipitation of virus and antibody using either agar gel or in liquid media.

15.2.1 Tube Precipitation Test

In this test narrow, thin walled glass tubes are subjected with purified virus suspension and antiserum and incubated at 37°C in a water bath. A precipitate is visible at the interface of viral antigen and antiserum. The type of precipitate will vary with type of antigen used. In case of elongated virus particles, floccular precipitate is formed while it is densely granular in case of spherical particles. It is one of most widely used test. For this test, relatively high concentrations of reagents are used and the reactants should be diluted in saline (0.85%).

15.3 Double Immuno-diffusion or Ouchterlony Test

Double diffusion technique can be performed either in Petridishes or on microscopic slides. It is a most widely used immunological technique which involves diffusion of both antigen and antibody move towards each other. For preparation of gel, agarmized in saline solution or in various buffers viz., citrate, phosphate, borate depending on the type of virus can be used. After agar has set in, wells are cut with the help of cork borer of 5-10 mm diameter. Antiserum is placed in central well and surrounded wells by antigens. After some time, diffusion of reactants start from their respective wells, continues towards each other and form complexes at their specific point of contact. If precipitate lines cross or intersect each other without mutual interference indicates the presence of two unrelated or distantly related antigens. Complete fusion of precipitate lines at their proximal ends indicates that two antigens form complexes with some antibodies so that two antigens are serologically related.

Materials required : Glass Petridishes, Sodium azide, cork borer, agarose gel, saline (0.85%), 0.07 M phosphate buffer pH 7, 0.6M sodium borate buffer pH 7.5.

Procedure : Pour an appropriate amount of agar in Petridishes. For this mix 0.6-1.0 gm of agar, 0.85% saline and 100 ml distilled water of phosphate buffer pH 7.0. After solidification, cut wells in gel with the help of a cork borer according to a predetermined pattern and the distance between wells should be 2-4 mm. Dilute antiserum with saline or PBS to prepare several dilutions, place antiserum in the centre well and antigen to the surrounding wells. Extract the sap in a minimum buffer, use healthy sap of the same plant as control. Incubate the dish at room temperature. In case of positive reaction, precipitation bands many appear between antigen and antibody.

15.4 Enzyme Linked Immunosorbent Assay

Application of Enzyme- linked immunosorbent assay (ELISA), a serological technique based on enzymatic amplification of the signal of antigen-antibody interaction for sensitive and reliable detection of plant viruses, was developed by Clark and Adams (1977). It is a solid phase procedure and is carried out in microtitre plates. In this technique, enzyme conjugated antibody bind with its homologous antigen and the antigen-antibody complex is probed by adding enzyme substrate, which upon enzymatic breakdown produces a colored product. The relative concentration of the antigen is determined by intensity of color which in turn is the function of antigen titre and time. It is also referred to as enzyme-immuno assays (EIA). For simplification and enhanced reliability several modifications of the ELISA technique has been attempted. The use of fluorogenic substrates (Torrance and Jones, 1982), modified antibodies F(ab)2-fragments (Adams and Barbara, 1982), enzyme amplification (Johannnson et al., 1985), enzyme amplification (Johannson et al., 1985; Torrance, 1987) or time-resolved fluorometry (Siitari and Kurppa, 1987) were some of the major modifications made during 1980s. Indirect and direct antigen-binding dot enzyme immunoassays (Dot EIA) on Nitrocellulose or nylon membrane, in place of the polystyrene solid phase of ELISA, were found to be very useful in comparison to the solid phase EIA (Rybicki and Von Wechmar, 1982).

These modifications of EIA can be grouped under two categories viz., direct and indirect ELISA. When primary antibody-enzyme conjugate binds directly to the antigen, it is known as direct ELISA while if primary antibody –enzyme conjugate does not bind directly to the antigen, it is referred to as indirect ELISA. Further DAS-ELISA is more specific and shows narrow spectrum i.e. may not detect distinct strains of the same virus. While, triple antibody sandwich (TAS-ELISA) is less specific, broad spectrum and more sensitive and hence useful in the establishment of serological relationship among strains of the same virus or between different viruses (Koenig and Paul, 1982). Another modification viz., Microarray-based multiplex ELISA has modernized and enhanced the efficacy of hitherto used serological assays and specifically detect numerous antigens simultaneously (Mendoza et al., 1999). Production of monoclonal antibodies (McAbs) of high affinity and specificity has increased the reliability of ELISA. Due to reasonably high sensitivity, reliability, specificity and amenability to automation, ELISA has become the most popular and useful techniques particularly for large-scale indexing of seed/ propagative stocks like potato and many other horticultural crops (Singh et al., 1989, 2000 a, b; Singh 2000).

There are two main categories of ELISA procedures namely Direct ELISA and Indirect ELISA.

The basic principle of the Direct ELISA lies in immobilizing the antigen onto a solid surface or captured by specific antibodies and probing with specific immunoglobulins carrying an enzyme label. The enzyme retained in the case of positive reaction is detected by adding the suitable substrate. The enzyme converts substrate to a product, which can be easily recognized by its color. Hence, ELISA is the amplification of reaction between viral antigens and their antibodies utilizing an enzyme and its substrate.

While in Indirect ELISA, Immunoglobulins (lgG) extracted from virus-specific antiserum are used for coating the solid surface to trap the antigen and the same immunoglobulins (lgG) labeled with an enzyme are employed for detection. In this case the antigen gets sandwiched between two immunoglobulins (lgG) and thus is referred to as the double – antibody sandwich (DAS) form of ELISA. It is highly strain specific. In direct ELISA no secondary antibody is involved. The antigen is trapped between one and the same antibodies specific to the virus raised in rabbit.

Limitations

This test is not suitable

- For virus detection in disease surveys unless it is targeted to a specific virus
- When adequate quantities of antisera are not available for lgG extraction and conjugation.
- For probing a single antigen with several different antisera.

15.5 Dot Immuno Binding Assay (DIBA)

The basic principle is similar to that of the plate ELISA. Antigens are first immobilized on nitrocellulose membranes, or nylon or poly vinylidine based membranes. In some cases, DIBA has been shown to be more sensitive than plate ELISA.

The main advantages of the DIBA over the plate ELISA

- It permits virus detection in extremely small volumes of test materials such as insect extracts.

- It is more suitable for virus detection in disease surveys because the membrane can be transported more easily than ELISA plates.
- Membranes are easier to process than ELISA plates.

(i)   Direct ELISA

Step 1	Step 2	Step 3	Type
Y	◊ over Y	E over ◊ over Y	DAS-ELISA
◊	E over ◊	–	DAC-ELISA

(ii)   Indirect ELISA

Step 1	Step 2	Step 3	Step 4	Type
Y	◊ over Y	◊ over Y	E over ◊ over Y	DAS-ELISA
◊	◊	E over ◊	–	DAC-ELISA

Chapter 16
Nucleic Acid Hybridization Techniques for Detection of Plant Pathogens

Nucleic acid hybridization is a well established tool in molecular biology, depends on high degree of specificity in nucleotide base pair sequences. Especially for diagnosis of plant pathogens, the technique is being applied because of its specificity. Different probes, such as radio active or non-radioactive probes are used for detection. One of the nucleotide sequences would be labeled with a probe that provides necessary signal required for hybridization. The most commonly used label is 32p which is incorporated into the detecting strand of nucleic acid. Using the nucleic acid probes, entire genetic information with in a pathogen can be studied.

16.1 Dot-blot Assay (DBA)

In this technique, the target nucleic acid from a pathogen or insect vector or plant sample is spotted onto a solid matrix. Nitrocellulose membranes are commonly used for the purpose and it is bound by baking. Using non-homologous DNA (salmon sperm or calf-thymus DNA) and a protein Bovine serum albumin or non-fat dried milk), the free binding sites on nitrocellulose membrane are blocked. After that hybridization is carried out with a labeled probe and being detected by autoradio radiography or calorimetric reaction. The procedure is similarly to ELISA though basis of binding is different. As the hybridization is based on annealing of nucleic acids of complementary strand while ELISA is based as binding of antigen and antibody. It is a recent recent development in hybridization technology offers great potential for detection of plant pathogen.

16.2 Polymerase Chain Reaction (PCR)

PCR is an ingenious new tool for molecular biology that has had effect on detection of plant pathogens similar to the discovery of restriction enzymes and the southern blot. In this technique a single DNA molecule can be amplified and single-copy genes are routinely extracted out of complex mixtures of genomic sequences that could be easily visualized as discrete bands on agarose gel. This method was invented by Kary Mullis

(1984), use of thermostable DNA polymerases and automation, have fostered the development of numerous and diverse PCR applications throughout he research community.

Performing a PCR

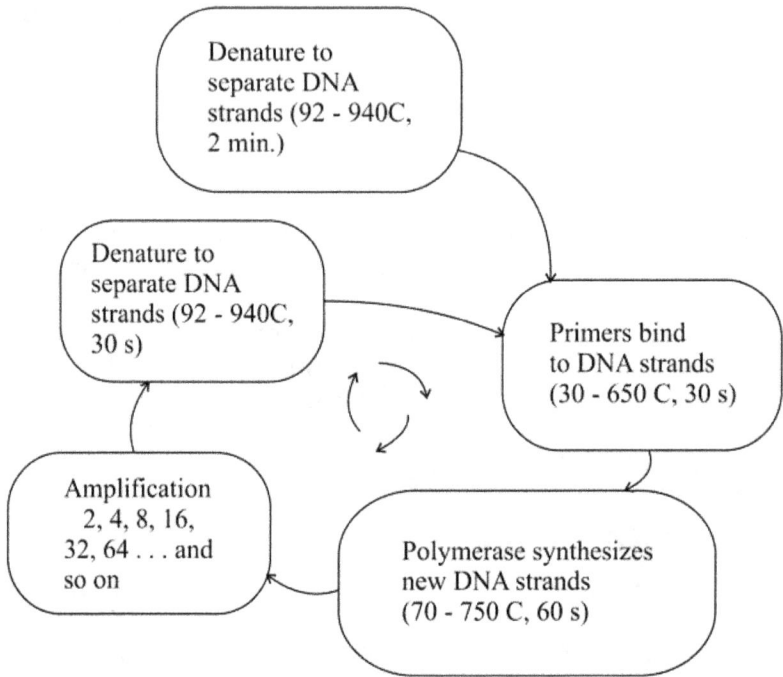

Denature to separate DNA strands (92 - 940C, 2 min.)

Denature to separate DNA strands (92 - 940C, 30 s)

Primers bind to DNA strands (30 - 650 C, 30 s)

Amplification 2, 4, 8, 16, 32, 64 . . . and so on

Polymerase synthesizes new DNA strands (70 - 750 C, 60 s)

Principle : It is an *in vitro* method of nucleic acid synthesis by which a particular segment of DNA can be amplified. It involves two oligonucleotide primers that flank the DNA fragment to be amplified and repeated cycles of heat denaturation of DNA, annealing of the primers to their complementary sequences and extension of the annealed primers with a thermostable DNA polymerase.

These primers hybridize to opposite strands of the target sequence and are oriented so that DNA synthesis by the polymerase proceeds across the region between the primers. Since, the extension products themselves are also complementary and capable of binding primers, successive cycles of amplification essentially double the amount of target DNA synthesized in the previous cycle, results in an exponential accumulation of specific target fragment.

The basic purpose of the reaction is to make a huge number of copies of a gene. For this, it is necessary to have enough starting template for sequencing. There are three major steps in a PCR, which are separated for 30 or 40 cycles. This is performed on an automated cycler, which can heat and cool the tubes with the reaction mixture in a very short time.

(i) **Denaturation :** Performed at 94^0C, during denaturation, the double stranded DNA melts down and form single stranded. All enzymatic reactions stop down such as extention from a previous cycle.

(ii) **Annealing :** It is done at 54^0C. The primers are moved around caused by Brownian motion. Hydrogen bonds are constantly formed and broken between the single stranded primer and the single stranded template. The more stable bonds last a little bit longer (primer that fit exactly) and on the little piece of double stranded DNA (template and primer), the polymerase can attach and starts copying the template. Once there are few bases built in, the hydrogen bond is so strong between the template and primer, that it does not break any more.

(iii) **Extension :** At 72^0C, which is the ideal temperature for the polymerase. The primers, where there are a few bases built in have a stronger attraction to the template created by hydrogen bonds than the forces breaking these attractions. Primers that are on positions with no exact match get loose again because of high temperature and donot give an extension of the fragment.

Protocol : While the standard conditions will amplify most target sequences, they are presented here principally to provide start-up or standard conditions for designing new PCR applications. It can be highly advantageous to optimize the PCR for a given application, especially repetitive diagnostic or analytical procedures in which optimal performance is necessary.

Set up a 25 μl reaction in a 0.5 ml microfuge tube, mix and overlay with 10 μl of mineral oil.

Template DNA	:	5 μl
PCR buffer	:	2.5 μl
25 mm $MgCl_2$	:	2.5 μl
5% Tween 20	:	2.5 μl
50 μm each dNTP	:	1 μl each
Primers	:	1μl each
Taq DNA polymerase	:	1-2 units

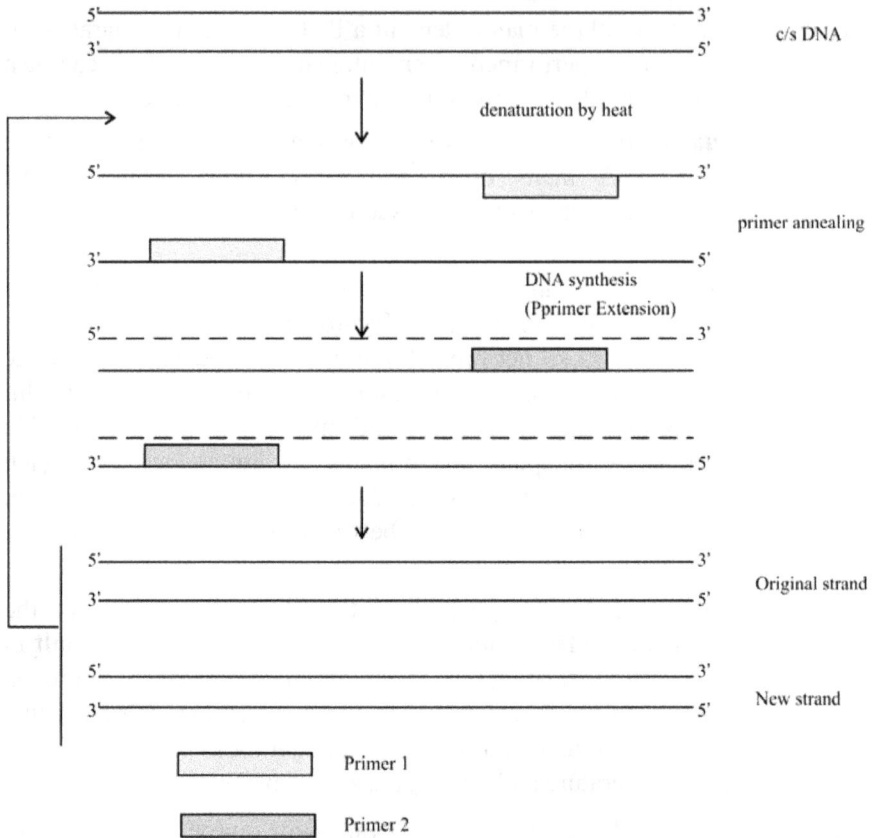

16.3 Extraction of DNA for PCR

For obtaining good quality of DNA, it is better to select infected plant leaves which are often rich in phenols and polysaccharides. Mainly C-TAB method is followed for extraction of nucleic acid.

Procedure : Weigh 100 mg of leaves and grind them with liquid nitrogen and transfer the powder to appendorf tubes. Add 700-1000 µl of C-TAB solution extraction buffer and vortex the contents to mix well for few minutes and incubate the reaction mixture for 30 min. at 65°C in water bath. Place the tubes in ice for 2 min. to homogenise and centrifuge at 14,000 rpm for 10 min. Then collect the supernatant and perform phenol chloroform extraction test and vortex it. Add 800 µl of chloroform : Iso-amyl alcohol (2:4:1) to this and centrifuge at 14,000 rpm for 10 min. Add 0.6 volume of ice-cold iso-propanol to supernatant and again centrifuge at 14,000 rpm per few minutes, keep at –20°C for overnight and centrifuge again. Then wash the pellet with 70% ethanol and centrifuge for 5 minutes.

Air dry the pellet and dissolves the pellet in TE buffer and store at -20^{0}C and collect the DNA obtained.

16.4 Considerations for PCR Optimization

(a) *Enzyme concentration :* The recommended Taq DNA polymerase in a 50 µl reaction is 1.25 units. Use of excess enzyme does not increase product yield significantly. It has also been observed that increased amount of enzyme and excessively long extension times increase the likelihood of generating artifacts resulting smearing in agarose gels.

(b) *Primer design :* PCR primers generally range from 15-30 bases in length and G+C content should be 40-60%. The sequence of the primers car also include regions at the 5' end for downstream applications care should be taken to avoid sequences which would produce internal secondary structure, ideally both primers should anneal at the same temperature. The 3' ends of primers should not be complementary to avoid primer-dimers. 50 pmol of primer (1 µm final concentration in a 50 µl reaction) is recommended as a starting point for optimization. Regardless of primer choice, the final concentration of the primer in the reaction must be optimized.

(c) *Magnesium Concentration :* It is a crucial factor affecting the performance of Taq DNA polymerase. The reaction components and chelating agents (eg: EDTA or Citrate) all affect the amount of free magnesium. Taq DNA polymerase will be inactive in the absence of adequate free magnesium while excess free magnesium reduces enzyme fidelity and may increase the level of non-specific amplifications. For this reason, it is important to determine the optimum Mgcl$_2$ concentration empirically for each reaction series containing 1.0-3.0 mm Mg^{2+} in 0.5mm increments by adding 2,3,4,5 or 5 µl of 25 mm Mgcl2 stock to 50 µl reaction.

(d) *Buffer considerations :* The enzyme Taq DNA polymerase requires the presence of a 10x reaction buffer, a critical component of it being Triton x-100. There will be little or no activity of enzyme results use of a buffer sans Triton x-100.

(e) *Template considerations :* successful amplification of the region of interest is dependent upon the amount and quality of template DNA. The amount of template required for successful amplification is dependent on the complexity of DNA sample. To start with, 10^4 copies of target sequence to obtain a signal in 25-30 cycles with a final DNA concentration of the reaction at <10ng/ µl.

(f) *Nucleic acid cross contamination :* It is important to take great care to minimize the potential for cross-contamination between samples and to prevent carryover of DNA and RNA from one experiment to the next. Separate work areas and pipettors for pre and post amplification step, positive displacement pipettes, gloves and other sterilization techniques ought to be used to prevent cross contamination.

(g) *Cycle parameters :* In order to determine melting temperature (Tm) of nucleic acid, many formulas exist. It is best to optimize the annealing conditions by performing the reaction at several temperatures, starting 5^0C below any calculated Tm approximately. The formula used to estimate the melting temperature for oligonucleotides is

$$Tm = 81.5 + 16.5 \ (\log_{10} [Na^+] \ 0.41 \ (G + C) -675/n$$

Where $[Na^+]$ is the molar salt concentration.

N = number of bases in the oligonucleotide

If smear is seen, lower the template DNA concentration (10-20µg), adjust $MgCl_2$ concentration or adjust annealing temperature.

Analysis of PCR products has been carried by agarose gel electrophoresis with molecular weight marker. The bands of DNA at expected base pair length will be very prominent indicating amplification. By comparing with molecular weight markers, the approximate length of bands in base pair can be estimated and the polymorphism exhibited by different samples can be scored. Using gel documentation system, the gel can be photographed.

16.5 Amplified Fragment Length Polymorphism (AFLP)

It is a diagnostic tool often used in molecular biology, allows differentiation between individuals, genotypes and strains. It is used in assessing phylogeny and genetically diversifying genes in plants (Vos et. al 1995), as it has characteristics of an ideal system for detecting genetic variation. It is based on amplification of subsets of genomic restriction fragments using PCR. With the help of restriction enzymes DNA is cut at specific sites and double stranded adapters are ligated to the end of the DNA fragments to generate DNA template for amplification. For amplification of restriction fragments, the sequence of adapters and adjacent restriction site serve as primary binding sites. Those restriction fragments in which nucleotides flanking the restriction site match the selective nucleotides during PCR. The fingerprints are generated by analyzing the subset of amplified fragments by denaturing PAGE.

It is a highly sensitive method for detecting polymorphism throughout genome. It is more robust and reliable than RAPD's because more stringent reaction conditions are used.

Procedure

- With the help of restriction enzymes, DNA is cut, twice.
- Ends of DNA fragments are being with double stranded oligonucleotide adopters.
- Amplification of sets of restriction fragments selectively.
- Restriction fragments are carried out with 32p labled probes / primers according to sequence.
- Gel analysis of amplified fragments.

16.6 Multiplex – PCR

In the recent past, molecular techniques are gaining importance in plant diagnostic work, enabling a sensitive detection of plant pathogens necessary for quarantine certification programmes. However, most of these systems are based on detection of only one pathogens. It is usual that crop being affected by more than one pathogen or mixed infections. In commercially important crops like chilli, potato often becomes infected with two to three different viruses. At present, the method used for detection of viruses at large scale is ELISA, and there are no procedures to detect mixed in infections in one step reaction. Using multiplex – PCR, it has been possible to detect mixed infections of virus pathogens in crop plants. For example the potato crop is being affected with potato virus Y (PVY, potyvirus) Potato virus x (PVX, Potex virus) and potato leafroll virus (PLRV, Luteovirus). Multiplex RT-PCR for PVY strains has been reported by Nei and Singh (2007) Multiplex PCR is increasingly used because it improves efficiency of diagnostic PCR.. It could be adapted for simultaneous detection of viruses of one particular crop and also for the detection of other major plant pathogens, such as fungi, bacteria, viruses and viroids in the same reaction simultaneously.

16.7 Restriction Fragment Length Polymorphism (RFLP)

It is a non-PCR based technique used to produce a finger print based on variation in the sequences, which can be used for identification and hence, it is otherwise called as "DNA finger printing". By using a specific combination of restriction enzyme and probe, the pattern of radioactive bands produced from southern blots will be very unique or specific to a particular species, strain or individual.

Basic steps involved are

- Isolation of DNA
- Using restriction enzymes, the DNA is cut into smaller fragments
- DNA fragments are separated by using Gel Electrophoresis.
- Transfer of DNA fragments to a Nylon or Nitrocellulose membrane.
- Using radioactive labeled probes, specific DNA fragments are visualized.
- Autoradiography and analysis of results.

Chapter 17
Tissue Culture Studies

17.1 Landmark Events

The phenomenon of totipotency of cells i.e the potentiality of cells to regenerate a whole organism from a single cell as given by **Haberlandt** (1902), laid the foundation stone of plant tissues and cell culture and was justifiably recognized as **"Father of Plant Tissue Culture".** The concept of tissue culture is initially based on the idea that any organ of the plant body can be fragmented into smaller parts termed as ' **explant'** and each cell of the callus can be developed into a whole plant. The first major steps into experimental plant tissue culture were made by Kotte (1922) using intact root tips of pea and maize on various nutrient media. The basic nutrient media used in most of the cultures are various modifications of formulations developed by Gautheret (1934) and White (1936), the pioneering scientists in the field of plant tissue culture, who demonstrated that tissues could be grown continuously in culture media. A standard nutritive medium consists of a balanced mixture of macro and micronutrients, vitamins, organic growth factors and plant growth regulators. Muir et al. (1946) obtained **suspension cultures** by transferring fragments of callus tissue to agitated liquid medium and succeeded in obtaining suspensions containing single cells and cell clumps. Nickell (1956) achieved continuous growth of cells in suspension culture of a variety of French bean (*Phaseolus vulgaris*). However, the establishment of totipotency of higher plant cell was not realized till the discovery of hormonal regulations by **Skoog** and **Miller** (1957). They studied the response of plant tissues to hormones in tobacco callus cultures and the synergistic effect of auxins and cytokinins in promoting cell division. The studies of Skoog and Miller suggested that the balance between cytokinins and auxins determines whether unorganized growth of tissues or shoot or root development. They observed that a relatively higher ratio of cytokinins to auxins results in **shoot differentiation** and the reverse favours **root differentiation**. Further, the development of experimental procedures reported for culture of tissues (Reinert, 1959), cell suspensions (Steward et al.1958) and eventually single cells (Vasil and Haberlandt, 1965) leading to *in vitro* regeneration of whole plants.

Effect of gibberellic acid (GA3) on the growth and metabolism of plants was reported by Bergmann (1960). The promotory or inhibitory effect of GA_3 on organogenesis might be due to its influence on auxin metabolism as reported by Kogal and Elma (1960).

Murashige (1961) found that exogenous supply of gibberellins in the presence of auxins and cytokinins was found to express the growth of tobacco and rice calli. For establishing tissue cultures, auxins have been proved to be an essential supplement. Where as, cytokinins control cell division and cell growth by exerting influence on specific proteins and nucleic acid metabolism. Establishment of callus growth with subsequent organogenesis has been obtained from many species cultured *in vitro* and factors controlling growth and differentiation in cultures were studied by Vasil and Vasil (1972) and Murashige (1971). For efficient morphogenesis, various growth regulating stimuli should be applied not only in right amounts but also in right sequence under acurate cultural conditions (Gamborg et al. 1974).

In addition to growth hormones, physical environment plays an important role in organogenesis, the major factors being the physical state of the medium, light, temperature, humidity as well as size, age and source of the explant etc. and cytokinin, in combination with an auxin was reported to be essential for growth and induction of somatic embryogenesis (Fujmura and Kominine, 1980). The science of plant tissue culture has taken a new direction with the discovery of extra chromosomal genetic element with autonomous replication, **plasmids,** from a Gram-negative soil bacterium *Agrobacterium tumefaciens,* which contains T_i plasmid (Tumour-inducing) capable of transferring genetic characters to the desired genotype. With this development the science " Plant Biotechnology" has emerged, though not new, to accommodate new areas of research. Thus, plant tissue culture and genetic engineering form an integral part of plant biotechnology.

17.2 Tissue Culture in Plant Pathogen Interaction

In the recent past, study and development of disease resistant plant through tissue culture technique has gained importance to augment and supplement the effort in the conventional and established diseased resistance breeding systems. With the advent of different biotechnological tools available in the last one decade, development of a disease resistant variety through tissue culture has proved to be useful technique to study the degree of susceptibility or resistance to microorganisms including plant pathogenic fungi.

17.2.1   Plant Material and Callus Initiaiton

Any part of the plant organ or body can be used for tissue culture studies called as **explant.** Callus is unorganized mass of cells initiated from disorganized growth of plant tissues of explant grown on a suitable nutritive medium. Explants used are apical meristem tip, shoot, root, leaf, flower, bud, another and pollen, axillary bud, mature and immature seeds, cotyledons, hypocotyl and epicotyl of young seedlings etc.

17.2.2   Morphological Features of Callus

A perceptible variation in the texture and appearance of callus cells can be observed, ranging from hard nodular mass to soft friable cell clumps. The shape of cells varied markedly from spherical to enlarged oblong shaped cells. It almost resembles ' **cauliflower like'** growth on the basal nutritive medium. The colour of callus may be white, cream white, green either in whole or part due to development of chloroplast or purple colour due to **anthocyanin** pigmentation.

17.2.3   Callusing Frequency

The frequency of callus initiation can be measured as the ratio of number of explants showing callus initiation to the total number of explants inoculated and expressed in percentage.

Callusing frequency = (No. of explants induced callus /

Total no. of explants plated) × 100

The callus produced from the original explant is called **primary callus**. The callus initiated from cells or tissues fragmented from primary callus is said to be **secondary callus**. These cultures can be maintained successfully for indefinite period by periodic subculturing. The time interval starting from callus initiation to till its transfer is called **passage**. Significantly, three stages of development are evident during initiation of unorganized callus from a new explant such as induction of cell division, a period of active cell division and become de-differentiated, cessation of cellular division and increasing cellular differentiation.

17.2.4   Signficance of Callus Cultures

(a) It forms the starting material for treatment with fungal culture filtrates or crude toxins.

(b) Essential for suspension culture through which cells are separated.

(c) It forms basis for vegetative propagation of plants.

17.3 Suspension Culture

Akin to callus cultures, cell suspension cultures are derived from any part of plant. But the usual practice of suspension cultures initiation has been from the callus already growing in culture. It is the culturing of isolated cells in liquid media and achieved by transferring the callus from original explant to the nutritive medium that supports the growth of friable callus and subjected to constant agitation by means of orbital shaker at 100-200 rpm. This is done to ensure uniform distribution of cells of various sizes and cell clumps in the medium, break up the cell aggregates and facilitates gas exchange to sustain cell respiration in the medium.

Cell multiplication in suspension culturing typically follow characteristic growth curve as in the case of bacteria, when inoculated into a culture medium. Cell division progress at a faster rate in suspension culture than in callus cultures. The four stages of growth cycle are

(i) *Lag phase :* when the cells are inoculated into a culture medium, the cells tries to adapt to replenished supplies of nutrients and at the end of lag phase, cells undertake all necessary synthesis prior to cell division. During this phase, there is little growth in cell number.

(ii) *Log phase / Exponential phase :* Cell division progresses at a faster rate, results in exponential increase in cell number and cell mass. Growth multiplication takes place as long as nutrients are present in the medium.

(iii) *Stationary phase :* Soon after the exhaustion of nutrients, the rate of cell division within the culture decreases and finally ceases. Accumulation of toxic meatabolites starts, results in death of cells.

Suspension cultures need to be transferred during exponential phase for optimum growth. For maintenance of suspension cultures, single cells or single clumps are to be transferred onto a fresh medium. Suspension cultures can be utilized as

(i) to study the effect of fungal culture filtrates or chemicals,

(ii) to develop specific cell line

(iii) for commercial production of cell mass.

17.4 Treatment of Suspension Culture with Fungal Culture Filtrate

The degree of susceptibility and resistant levels of genotypes can be studied by treating the cells derived from suspension culture with fungal culture filtrates. For this purpose, healthy looking calli is to be taken into an embryo dish containing 10 ml of crude toxin. Immediately, a small amount of toluene is to be added to the dish (containing callus pieces) so as to form a thin impermeable layer to avoid from external infection. The calli are periodically observed under a binocular research microscope at regular intervals to look for cellular changes. After specified period of treatment, when cellular changes in calli are observed, the calli can be removed, wash with distilled water and use for regeneration ability studies and also for electrophoretic studies. The cellular changes can be depicted in the form of **plasmolysis**.

The degree of plasmolysis is calculated by using ocular micrometer. The percent of plasmolysis can be calculated by using formula

$$P = (Cc - C_t/C_c) \times 100$$

where,

P = percent of plasmolysis

Cc = diameter of callus cell in control

C_t = diameter of callus cell in treatment

17.4.1 Morphogenetic Studies

The suspension culture cells treated with fungal culture filtrate will be subjected to transformation studies, to evaluate the totipotency of an individual cell transformed into a whole plant. It will be carried with basal nutritive medium supplemented with high kinetin and low auxin content.

17.5 Isozyme and Protein studies

Study of isozyme bands and protein profile form a powerful biotechnological tool to study the phenotypically indistinguishable genetic systems. The traditional markers used in genetic analysis are quite often based on phenotypes, morphology, colour and other qualitative and metric traits. Isozymes and proteins help in the study of the interaction of naturally occurring allelic products and differential gene action in the development and gene expression. Isozymes and proteins provide a natural built-in-marker system for varietal identification, study of growth and differentiation in tissue culture, biochemical characterization and genetic analysis.

To characterize germplasm resources, biochemical and molecular techniques such as electrophoresis to soluble proteins and enzymes provide discrete genetic marker loci. Proteins and enzymes are charged molecules capable of mobility in an electric field. The charge of proteins is a function of pH. Based on this principle, the electrophoresis technique has been developed, which is a most powerful analytical tool available to separate isozymes and proteins. In other words, the presence of specific enzymes or proteins in a given plant material as evidenced by the electrophoresis can be directly correlated to the activity of specific genes. The proteins and enzymes have a pH dependent charge on their surface.

The term isozyme was coined by Markert and Moller (1959) to refer to the occurrence of multiple molecular form of an enzyme capable of catalyzing the same reaction. It may be coded by different alleles at the same loci or different loci within the same organism. According to the current recommendations of the commission on biological nomenclature of IUPAC-IUB, Isozymes are defined as "multiple molecular forms of an enzyme occurring within a single species as a result of presence of more than one structural gene". Multiple forms of protein may be due to the presence of multiple gene loci of multiple alleles.

Polyacrylamide gels are used as the supporting medium for seperation of enzymes and protein fractions (Raymond and Weintraub, 1959). Sodium-dodecyl – Sulphate (SDS) PAGE and Non-SDS PAGE will be used to study proteins and Isozymes, respectively.

17.5.1   Preparation of Zymograms.

The zymograms will be prepared and quantified by calculating relative mobility of protein and isozyme bands.

$$\text{Relative mobility } (R_m) = \frac{\text{Distance traveled by protein } \backslash \text{ enzyme front}}{\text{Distance traveled by dye front}}$$

The electrophoretic patterns can be analyzed qualitatively and quantitatively. The qualitative description can be given in the form of photographs, zymogram drawings with relative mobility values. Where as the quantitative analysis can be done in the form of differentiation through coefficient of similarity values, comparison of similar bands and hypergeometric distribution technique.

17.6 Hypergeometric Quantification of Electrophoretic Bands

Hypergeometric distribution occupies a place of great significance in statistical theory. It applies to sampling without replacement from a finite population. This can be traditionally expressed in the following form (Steel and Torrie, 1930).

$$P(n_1) = \frac{n_1^{N_1} - n - n_1^{N-n_1}}{n^N}$$

$$n_1^{N1} = N_1! / n! (N_1 - n_1)!$$

This is modified to suit the quantification of electrophoretic bands by Leibermann and Owen (1961) in the following form

$$P(x) = K! \; n! \; (N-K)! \; (N-n)! \; / \; (K-X)! \; (n-x)! \; X! \; N! \; (N-K-n+x)!$$

where,

$P(x)$ = Probability of obtaining chance or accidental matches.

K = Number of bands observed for one genotype / isolate

n = Number of bands observed for other genotype / isolate

N = Total number of bands observed for both genotypes / isolates

X = Number of matches or matching bands.

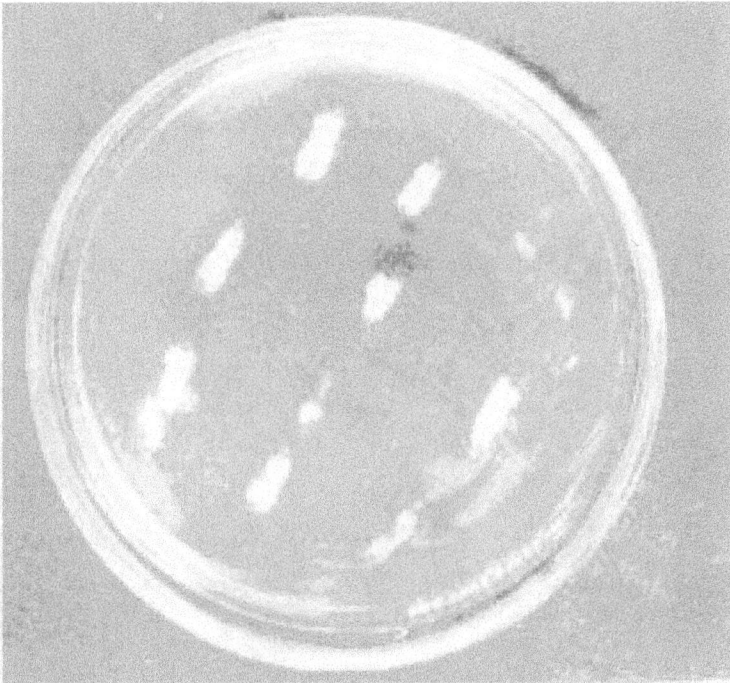

Fig. 17.1   Callus initiation from apical meristem explant of blackgram genotype LBG-22.

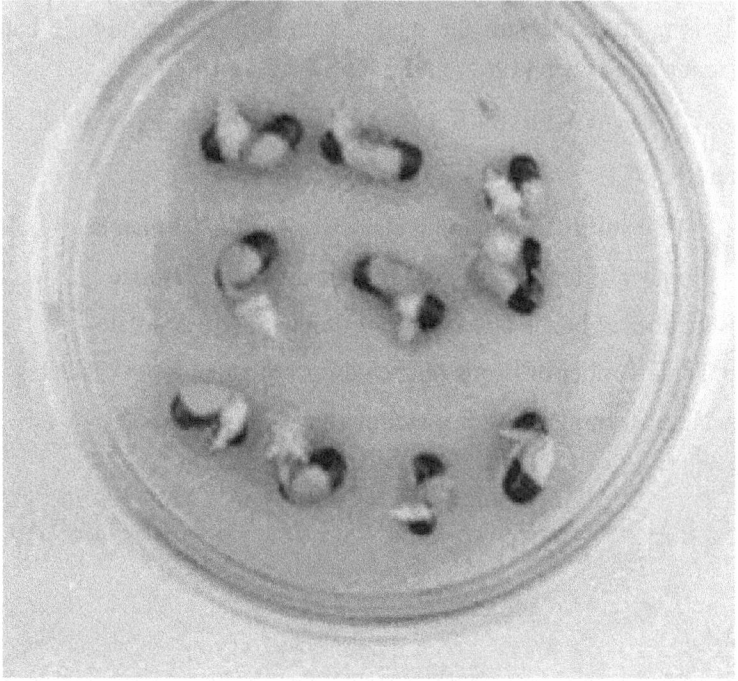

Fig. 17.2 Callus initiation from hypocotyl region of blackgram genotype LBG-22.

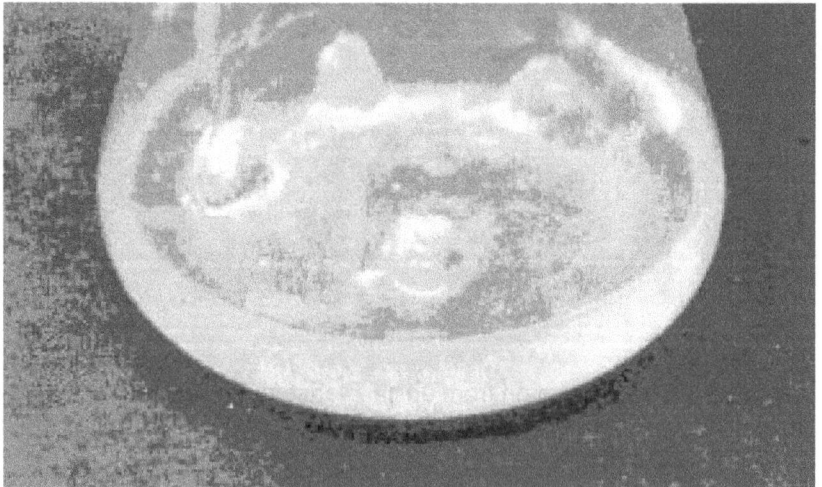

Fig. 17.3 Subculturing and maintenance of callus.

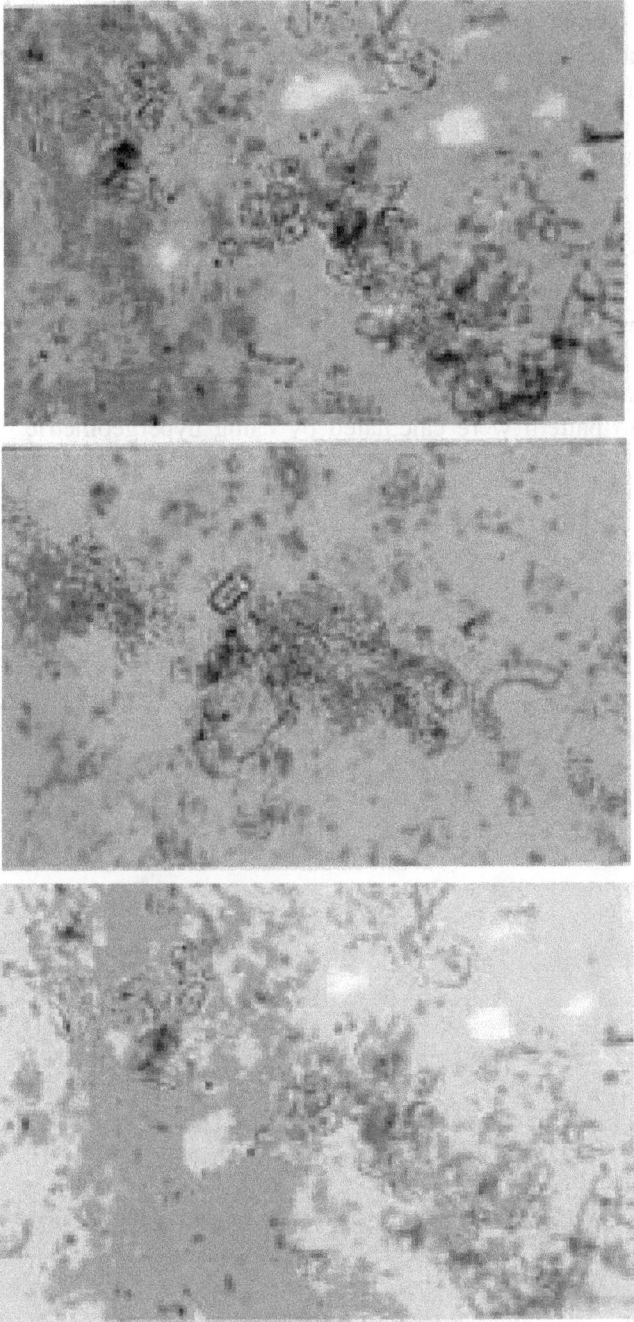

Fig. 17.4 Callus cells treated with fungal culture filtrate of *Corynespora cassiicola*, incitant of Blackgram leafspot showing varying degrees of plasmolysis.

Reddi Kumar and Balasubramanian (2007) studied the influence of auxins and cytoxivins on callus induction from different explants of blackgram genotypes. Murashige and Skoog (MS medium) medium supplemented with 3 mg/lt NAA and kinetic mg/lt showed the highest callus induction from apical meristem tip explant of different blackgram genotypes. The calli derived from apical meristem of LBG-648 showed more number of viable cells after treating with undialysed and dialysed culture filtrates of *Corynespora cassicola*. The coefficient of similarity values for all genotype combinations of stem tip callus treated with culture filtrate were found to be minimum (<0.5) when isozymes viz; polyphenol oxidases, esterases and proteins were studied. The genotypes III and V LPG-22 and LBG-648 were found to be closely related when the isozymes and protein pattern were calculated by using hypergeometric distribution technique.

Chapter 18

Genetic Engineering
Transgenic Plants

Among the diseases caused by plant pathogens, management of viral diseases poses a difficult task because of complex disease cycle, efficient transmission of vector and no effective viricide so far. Viruses are ultramicroscopic microorganisms majorily composed of proteins and nucleic acids, the genomic component being either DNA or RNA. Today more than 1000 plant viruses are known in varied form, induce diseases on plants. Nearly 70 groups of viruses belonging to 17 families are known to infect plants (Van Rogenmortel et. al 2000). Among viral groups, the largest number of diseases are causes are caused by poty viruses and gemini viruses. Other groups such as ilar, topso and badna groups are fast emerging as serious pathogens, the nucleic acid of viruses is ss RNA, dsRNA, ssDNA and ds DNA, which is encapsidated by protein subunits in regular pattern.

Integration of various approaches for management of diseases include cultural practices, control of vectors, avoidance of source of infection and use of resistant host plants. The latter one seems to be the most practical approach in containing the disease. But, due to lack of source of resistant germplasm and rapid development of resistance breaking strains of viruses make development of resistant varieties very difficult. Genetic engineering plays a pivotal role to overcome the difficulties associated with the conventional breeding in developing new varieties with durable resistance by pyramiding the genetically engineered resistant genes.

18.1 Sources of Transgenes for Protection Against Viruses

- Pathogen – derived resistance (PDR)
- Natural resistance genes
- Genes from other sources

18.2 Pathogen Derived Resistance

Hamilton (1980) and Sanford and Johnson (1985) mooted the concept of pathogen-derived resistance (PDR). They suggested that transgenic expression of pathogen sequences might interfere with the pathogen it self. Powell – Abel et. al., (1986) who first demonstrated PDR against plant viruses and protected tobacco plants from infection by expression of TMV coat protein. Since then many different viral genes and viral associated RNA's have been used as transgenes to confer resistance in plants and especially against a range of plant viruses having positive sense ssRNA, ambisense RNA. (Hull, 2002; Varma et. al., 2001, Das Gupta et. al., 2003). Coat protein (CP) gene is the most commonly used transgene for developing transgenic plants followed by replicase protein (RP) and movement protein (MP) genes have been developed for a large number of crops against the viruses. Both coding and non-coding regions of viral genomes have been used for developing pathogen – derived resistance. Viruses depend on the host machinery for its application. Most of the plant viruses have positive sense ssRNA genomes, that replicate by virus encoded RNA dependent RNA polymerase and form dsRNA replicative intermediate (Varma and Ramachandran, 2001).

Events that follow infection are

- Disassembly of virus particles.
- Synthesis or transcription of mRNA
- Translation of protein coded viral genome for various functions.
- Maturation of particles.
- Vector transmission.

18.2.1   Coat Protein Mediated Virus Resistance

Coat protein (CP) – mediated resistance (CP-MR) refers to resistance of transgenic plants that produce virus coat protein to infections against the virus from which coat protein (CP) is taken. Gene encode for virus coat protein or capsid protein is the most widely used transgene to generate virus resistant transgenic plants. This is because coat protein is relatively easy to identify and clone. It also provide protection against other viruses of the same virus group. CP-MR has been found effective irrespective of virus particle morphology (rod or isometric), genome (positive or negative sense and mono or multipartite) and mode of transmission (mechanical, seed, leaf hopper, whiteflies, thrips, aphids).

Mechanism : Two types of mechanisms exist

    (a)   Protein mediated

    (b)   RNA mediated

The mechanism of resistance induced by CP gene is either through the protein encoded by the transgene (protein mediated) or by the transcript of the transgene (RNA mediated) (Lomonosoff, 1995; Reinmann – Philipp, 1998) or both. The protein inhibits disassembly of the infecting virus in the case of protein mediated resistance, RNA mediated resistance is of high level and virus specific and is attributed to lower levels of transcripts. CP-MR in many virus host-combinations is caused by post transcriptional gene silencing (PTGS). This plant defense system results in degradation of mRNA produced both by the transgene and the virus (Water house et. al., 2001).

Examples

(a) Plants transformed with CP gene of TMV is effective against Tomato mosaic virus (TMV).

(b) CP gene of Soybean mosaic virus is effective against Tobacco Etch virus (TEV) and PVY (NeJidat and Beachy, 1990).

Transgenic crop varieties commercialized based on CP-MR are

- Squash variety Freedom II is resistant to Watermelon mosaic virus -2 (WMV-2) and Zucchini yellow mosaic virus (ZYMV) (Tricoli et. al., 1995).

- Squash var. ZW-3 is resistant to Cucumber mosaic virus (CMV) WMV-2 and ZYMV (Fuchs et al., 1997).

- Papaya var. Sun up and Rainbow UH are resistant to Papaya ring spot virus (PRSV) (Gonsalves. 1998).

18.2.2 Replicase Protein

Replicase Protein (RP) gene is the second widely used transgene to confer resistance against the plant viruses.

Salient features

- High degree of specificity – Resistance is shown only to the virus from which the replicase gene was derived and to very closely related strains or mutants.

- Degree of resistance shown both to viral and to RNA inoculum.

Golemboski et al., 1990 first showed that viral replicase expressing proteins were offering protection against TMV in tobacco. This phenomena is referred to as replicase protein mediated resistance (RP-MR). In most cases, use of replicase sequences give RNA-mediated protection. RP-MR gives nearly immune type reaction and there is a substantial inhibition of virus replication in initially inoculated cells and some cell to cell movement, but the infection does not spread from the inoculated leaf and no systemic disease develops.

18.2.3   Movement Protein

Movement proteins are responsible for transfer of viruses from cell to cell, which are synthesized by viruses. It is an ideal strategy for developing transgenic resistance by interfering with cell to cell movement. This could be brought about by expressing a defective movement protein conferring resistance against viruses and it has been demonstrated for atleast six groups of viruses (Cooper et al. 1995; Hull, 2002). Unlike other strategies, this approach offers attractive possibility to confer broad-spectrum resistance to related and unrelated viruses.

Examples

(a) Defective movement protein of TMV expressed in transgenic tobacco plants was shown to confer resistance against viruses including Alfalfa mosaic virus (AMV), Tobacco rattle virus (TRV), Tobacco ring spot virus (TRSV) and CMV (Cooper et. al; 1995).

(b) Movement protein of mutant potato leaf roll virus (PLRV) exhibited resistance against unrelated viruses PVY and PVX in potato.

The other strategies such as Antisense RNA, Satellite RNA's can also be utilized in development of transgenic resistance against viruses.

18.3 Natural or Plant Derived Resistance Genes

In this case, a resistant gene from a plant can be isolated and readily transferred to another plant species for control of virus diseases.

Examples

Bendahmane et. al., 1999 isolated Rx1 gene from potato which gives extreme resistance to potato virus x (PVX) and transformed into *Nicotiana tabacum* and *N.benthamiana* where it offers resistance to virus. Likewise, the N gene confers resistance to Tobacco mosaic virus was isolated from Tobacco (*N.glutinosa*) showed resistance when transferred to tomato against TMV.

18.4 Genes from Other Sources

(a) *Antiviral proteins :* These have been identified from various plant species and genes coding them and transferred to other plants. Many of the antiviral proteins isolated are fall into the category of Ribose inactivating proteins (RIP's).

Example

When tobacco and potato plants are transformed with antiviral protein gene exhibited resistance against viruses like PVX, PVY, TMV (Lodge et al. 1993; Lodge et al. 1998)

(b) *Pathogenesis related proteins (PR Proteins) :* These are the proteins induced in plants following infection with viruses causing necrotic local lesions. For example, treatment of leaves with salicylic acid induces certain PR proteins and inhibits Alfalfa mosaic virus (Amv) replication. These are the proteins involved in local acquired resistance which are part of a non-specific host defense reaction. Using this strategy, it is possible to provide protection against certain viruses by using transgenic plants in which PR protein gene is expressed.

18.5 Problems Associated with Transgenics

Some of the potential risks involved in the use of transgenic plants resistant to viruses are

- Toxicity of gene products
- Adverse effect of marker genes like antibiotic resistance genes.
- Development of resistance breaking strains of virus.
- Chances of transfer of related genes to other species.
- Emergence of new recombinant viruses which may be more virulent and have altered host range.

Table 18.1 Resistance genes (R genes) against viruses in Plants

Virus	Resistance gene	Source plant
Turnip crinkle virus	RRT	Arabidopsis *thaliana* ecotype Dijon
Potato virus x	Rx, Nx, Nb	*Solanum tuberosum* cv. Cara
Tobacco mosaic virus	Ry	*Solanum stoloniferum*
Tobacco mosaic virus	Tm1	Lycopersicon *esculentum*
Turnip mosaic virus	TURBO1	*Brassica napus*
Tobacco veinal mottle virus	Va	*Nicotiana tabacum* cv. Burley
Pepper mild mottle virus	L_2	*Capsicum* sp.
Bean common mosaic virus	I	*Phaseolus vulgaris*

Table **18.2** Transgenic plants developed in various crops against viruses through coat-protein mediated resistance.

I. Cereals	
1. Rice	Rice tungro spherical virus, Rice stripe virus
2. Wheat	Wheat streak mosaic virus
3. Maize	Maize dwarf mosaic virus, maize chlorotic mottle virus.
II. Fruit crops	
1. Citrus	Citrus tristeza virus
2. Grape	Grapevine fan leaf virus, Grapevine chrome mosaic virus
3. Papaya	Papaya ring spot virus
4. Apricot	Plum pox virus
5. Squash	Zucchini yellow mosaic virus, cucumber mosaic virus, water melon mosaic virus -2.
III. Vegetables	
1. Tomato	Tomato mosaic virus, Tomato yellow mosaic virus, cucumber mosaic virus, Tomato yellow leaf curl.
2. Bean	Bean pod mottle virus
3. Pea	Pea enation mosaic virus
4. Potato	Potato virus x, Potato virus % Potato leaf roll virus.

(***Source:*** Das Gupta et. al. 2003)

Chapter 19

Nanotechnology for Detection of Plant Pathogens

Nanotechnology is a rapidly growing field of technology that produces new products and enhances the quality of human lives. It deals with the understanding and control of matter at dimensions of roughly 1 to 100 nanometers, where unique phenomena enable novel applications. At this level, the physical, chemical and biological, properties of materials differ in fundamental and valuable ways from the properties of atoms and molecules or bulk mater.

Nanotechnology is the manipulation or self – assembly of individual atoms, molecules or molecular clusters into structures to create materials and devices with new or vastly different properties. Nanotechnology allows the manufacturing and manipulation of matter at basically any scale, ranging from single atoms and molecules to micro-meter sized objects and enables the miniaturization of many current devices, resulting in faster operation or the integration of several operations. The term 'Nano' is devised from a Greek word means 'Dwarf' that is small. It is the ability to manipulate material at molecular level driving out novel and exciting properties. When a material is reduced down to 100 nm (1 nanometer – 10^{-9} meter), its all properties including chemical, electrical, thermal, mechanical and optical are altered. In the history of Nanotechnology, the first mention of some of the distinguishing concepts on nanotechnology was in 1867 by James Clark Maxwell when he proposed an atom size device known as "Maxwells demon". In the later years, the topic was touched upon in **"There's plenty of room at the bottom"** a talk given by **Richard Feynman** at an American physical society meeting at Caltech December, 29, 1959. The term 'Nanotechnology' was defined by Professor Norio Taniguchi of Tokyo Science university as "Nanotechnology mainly consists of the processing of, separation, consolidation, and deformation of materials by one atom or one molecule". Nanoscience and Nanotechnology got started in the early 1980's with two major developments one being the birth of cluster science and the other, invention of the scanning tunneling microscope (STM). This development led to the discovery of fullerenes in 1986 and carbon nanotubes a few years later. In another development, the synthesis and properties of semiconduction nanocrystals was studied. This led to a fast increasing number of metal oxide nanoparticles of quantum dots.

The Nanotechnology initiative (NNI) definition of Nanotechnology as **"Materials and systems whose structures and components exhibit novel and significantly improved physical, chemical and biological properties, phenomena and processes due to their nano-scale size".** The technology used for synthesis, characterization of materials and devices that operate at nanoscale as well as exposing the impact of devices and materials on human life is called **"Nanobiotechnology"**.

Table 19.1 Comparisons of scale from macro to molecular.

Size (nm)	Examples	Terminology
0.1-0.5	Individual chemical bonds	Molecular / atomic
0.5-1.0	Small molecules, pores in Zeolites	Molecular
1-1000	Proteins, DNA, mesospores,	Nano Inorganic nanoparticles
10^3-10^4	Devices on a silicon chip,	Micro Living cells
>10^4	Normal bulk mater	Macro

There is an enormous interest in the synthesis of nanomaterials due to their unusual properties such as optical, chemical and electronic. It is gaining importance in areas such as catalysis, optics, biomedical sciences, mechanics, magnetics, and energy science. It is well known that many organisms can provide inorganic materials either intra or extracellularly. For example, unicellular organisms such as magnetotactic bacteria produce magnetite nanoparticles and diatoms synthesize siliceous materials. Multicellular organisms produce hard inorganic – organic composite materials such as bones, shells, and spicules using inorganic materials to build a complex structure. These biominerals are composite materials and consist of an inorganic and a special organic matrix viz., proteins, lipids, or polysaccharides, that controls the morphology of the inorganic compound. The surface layer of bacteria produce gypsum and calcium carbonate layers. Even though many biotechnological applications such as remediation of toxic metals employ microorganisms are recently found as possible eco-friendly nanofactories.

19.1 Relationship of Nanotechnology to Science and Engineering in Agriculture

Today, in Agriculture, if a plant or animal becomes infected with disease, it can be days, weeks or months before disease presence is detected by whole-organism symptoms. By that time infection may be widespread and entire herds / fields might need to be destroyed. Nanotechnology operates at the same scale as a virus or infecting particle, and thus holds the potential for very early detection and eradication. Nanotechnology holds out that "Smart" treatment delivery systems could be activated long before macro symptoms appear. For example, a smart treatment delivery system could be a miniature device implanted in an animal that samples saliva on regular basis. Long before a fever develops, the integrated sensing, monitoring and control system could detect the presence of disease and notify the former and activate a targeted treatment delivery system. Smart treatment delivery systems are envisioned for biology and bioactive systems such as drugs, pesticides, nutrients, probiotics and implantable cell bioreactors.

In agriculture, the fundamental life processes are explored through research in molecular and cellular biology. New tools for molecular and cellular biology are needed that are specifically designed for separation, identification and quantification of individual molecules. This is possible with nanotechnology and could permit broad advances in agricultural research, such as reproductive science and technology, conversion of agricultural and food wastes to energy and other useful by-products through enzymatic nano-bio processing, disease prevention and treatment in plants and animals. New materials that have special characteristics at the nanoscale could offer a tremendous break through for pathogen and contaminant detection. Materials that have self-assembly and self-healing properties can find a multitude of applications in agriculture. Packaging of food in self-healing containers could prevent food microbial contamination and facilitate food preservation, storage and distribution. Protection of environment through the reduction and conversion of agricultural materials into valuable products can be made easier by nanotechnology.

Protection of the environment through management of local and environmental emissions is another exciting area of agriculture that could benefit from nanotechnology. Agricultural crops must be protected against the invasions of wild animals, weeds, insect pests, fungal pathogens and the whimsical nature of the weather. Close daily scrutiny or "scouting" of crops for potential problems is critical to the health of the crop and also reduces the amount of pesticides needed.

Computerized control of environment over small enclosed parcels of land is known as **Controlled Environment Agriculture** (CEA). CEA technology, as it exists today in US, Europe and Japan, provides an excellent stage for the introduction of nanotechnology into plant production agriculture. With many of the monitoring and control systems already in place nanotechnological devices for CEA that provide "Scouting" capabilities could tremendously improve the growers ability to determine the best time of harvest for the crop, the health of the crop and question of food security such as microbial or chemical contamination of the crop.

19.2 Nanosensors for Pathogen and Contaminant Detection

At present, sensors provide an abundance of information about such parameters as temperature and weather data and data that provide information on land, air, sea transportation and counts other variables. Biological organisms also have the ability to sense the environment. Humans sense the environment through sight, touch, taste, smell and sound. For example, the human ear uses nanostructures to transduce the macro-motion of ear drum-induced fluid motion into a chemical or electrical signal. In living organisms sensors operate over a range of scales from the macro (ear drum vibrations) to the micro (nerve cells) to the nanoscale (molecules binding to receptors in our noses).

The exciting possibility of combining biology and nanoscale technology into sensors holds the potential of increased sensitivity and therefore a significantly reduced response – time to sense potential problems. For example a bioanalytical nanosensor that could detect a single virus particle long before the virus multiplies and long before the symptoms were evident in the plants and animals. Some areas of potential applications for bioanalytical nanosensors are

- Detection of pathogens
- Detection of contaminants
- Environmental characteristics (Light / Dark Hot / Cold, Wet/Dry)
- Heavy metals
- Particulates or allergens

19.2.1 Desirable characteristics of biosensors

- Should be small in size
- Rapid response
- Portable
- Specific, qualitative, reliable
- Accurate and results should be reproducible
- Robust and stable

19.3 Nanodevices for Smart Treatment Delivery Systems

Today, application of agricultural fertilizers, pesticides, antibiotics, probiotics and nutrients is typically by spray or drench application to soil or plants, or through feed or injection systems is either provided as "Preventive" treatment or is provided once the disease organism has multiplied and symptoms are evident in the plant or animal. Nanoscale devices are envisioned that would have the capability to detect and treat an infection, nutrient deficiency or other health problem long before symptoms were evident at the macro-scale. This type treatment could be targeted to the area affected.

Smart delivery systems for agriculture posses any combination of characteristics such as time – controlled, spatially targeted, self- regulated remotely regulated or multifunctional characteristics to avoid biological barriers to successful targeting.

19.4 Role of Microbes in Synthesis of Nanoparticles

Nanomaterials are at the leading edge of the rapidly developing field of Nanotechnology. The development of reliable experimental protocols by the synthesis of nanomaterials over a range of chemical compositions, sizes, and high dispersity is challenging issues in current nanotechnology. Biotechnology systems, masters of ambient condition chemistry, synthesize inorganic materials that are hierarchially organized from the nano to the macroscale. Recent studies on the use of microorganism in the synthesis of nanoparticles are a relatively new and exciting area of research with considerable potential for development.

19.4.1 Bacteria in Nanoparticle Synthesis

Among the microorganisms, prokaryotic bacteria have received the most attention in the area of biosynthesis of nanoparticles. Early studies reveal that *Bacillus subtilis* 168 is able to reduce Au^{3+} ions to produce octahedral gold particles of nanoscale dimensions (5-25nm) within bacterial cells by incubation of cells with gold chloride under ambient temperature and pressure conditions. Iron reducing bacteria *Shewanella algae* can reduce Au (III) ions in anaerobic environments. In the presence of *S.algae* and hydrogen gas, the Au ions are completely reduced, which results in the formation of 10-20 nm gold nanoparticles.

It is already established that silver is highly toxic to most microbial cells. Nonetheless, several bacterial strains are reported as silver resistant and may even accumulate silver at the cell wall to as much as 25% of the dry weight biomass, thus suggesting their use for the industrial recovery of silver from ore material. The silver resistant bacterial strain *Pseudomonas stutzeri* AG 259 accumulates silver nanoparticles, along with some silver sulfide, in the cell where particle size ranges from 35 to 46nm.

Bacteria not normally exposed to large concentrations of metal ions may also be used to grow nanoparticles. The exposure of *Lactobacillus* strains, which are present in buttermilk, to silver and gold ions resulted in the large scale production of metal nanoparticles within the bacterial cells.

In addition to gold and silver nanoparticles, there is much attention in the development of protocols for the synthesis of semiconductors (the so-called quantum dots) such as Cds, Zns and Pbs. These luminescent quantum dots are emerging as a new class of materials for biological detection and cell imaging.

19.4.2 Fungi in Nanoparticle Synthesis

The use of fungi in the synthesis of nanoparticles is a relatively recent addition to the list of microorganisms. The use of fungi is potentially exciting since they secrete large amounts of enzymes and are simpler to deal within the laboratory. However, genetic manipulation of eukaryotic organisms as a means of over expressing specific enzymes identified in nanomaterial synthesis would be much more difficult than that in prokaryotes.

An extensive screening process resulted to two genera, which when challenged with aqueous metal ions such as Au cl$_4^-$ and Ag$^+$, yielded large quantities of metal nanoparticles either extracellularly or intracellularly. The appearance of a distinctive purple color in the biomass of *Verticillium* after exposure to 10^{-4} MH Au Cl$_4$ solution indicates the formation of gold nanoparticles intracellularly and can clearly be seen in the Uv- visible absorption spectrum recorded from the gold-loaded biomass as a resonance at 550 nm. The exposure of *Verticillium* sp. to silver ions resulted to a similar intracellular growth of silver nanoparticles. From the application point of view, it would be imperative to harvest the metal nanoparticles formed within the fungal biomass. It is possible to release the intracellular silver and gold nanoparticles via ultrasound treatment of the biomass-nano particles composite or via reaction with suitable detergents.

Nanotechnology is an enabling technology that has the potential to revolutionize Agriculture and food systems. The Cooperative State Research Education and Extension Service (CSREES) has identified specific research areas in agriculture and food systems, several of which can directly benefit from research in nanotechnology. The research areas are

- Pathogen and contaminant detection
- Identity preservation and tracking
- Smart treatment delivery systems
- Smart systems integration for agriculture and food processing.
- Nanodevices for molecular and cellular biology.
- Nanoscale materials science and engineering
- Environmental issues and agricultural waste
- Education of the public and future work force.

19.4.3   Viruses as Nanoparticle

Recently plant viruses have attracted attention as biological templates for assembly of nanostructures and nano electronic circuits. Viruses are known to cause diseases in plants, animals and human beings and harmful effects associated with them are multitude. But researchers have found a new way to use the plant, animals or bacterial viruses to be used as nanoparticles based on their size lying in nanosclaes. The ability of viruses to self-assemble into nano scale particles of discrete size and definite geometry gives them potential utility in a variety of nano and biotechnological applications (Peabody, 2003). Because of their well characterized nano scale structures and ease of large – scale production, virus particles have become a popular tool for cancer nanotechnology researchers aiming to develop targeted anti tumour therapies.

19.5 Nanotechnology Research Institutes in India

- Jawaharlal Nehru Centre for Advanced Scientific Research, Bangalore.
- National Physical Laboratory, New Delhi.
- Solid State Physics Laboratory, New Delhi.
- National Chemical Laboratory, Pune.
- Indian Institute of Science, Bangalore.
- Amity Institute of Nanotechnology, Noida
- Banaras Hindu University, Varansi.
- Indian Institute of Technology at Kanpur, Delhi, Chennai, Guwahati, Kharag pur, Mumbai.

19.6 Nanotechnology Teaching Institutes

- Amity Institute of Nanotechnology, Noida
- Punjab University, Chandigarh
- GB Pant University of Agriculture & Technology, Pantnagar.
- Sanjay Institute of Engineering & Management, Mathura.
- Vellore Institute of Technology, Vellore.
- Hindustan Institute of Agriculture & Technology, Agra
- Institute of Technology & Management, Gwalior.

Annexures

I **Composition and Preparation of Culture media for Isolation of Specific Fungi**

   (i) **Selective media for isolation of *Paecilomyces lilacinus***

      (a) **Medium of Mitchell et al. (1987)**

Nacl	:	10 g
Pentachloronitro benzene	:	500 mg
Benomyl	:	0.5 mg
PDA	:	39 g
Deionized water	:	1 lit.
After autoclaving add		
Chlortetracycline hydrochloride :		50 mg
Tergitol	:	1 ml.

Serial dilution of soil and 1 ml of an appropriate dilution of soil-water suspension is pipetted into an empty sterile petridish containing selective medium and incubate at 25-27^{0}C.

   (b) **PSM Medium (Cabanillas and Barker, 1989)**

Streptomycin sulphate	:	50 mg
Chlortetracycline hydrochloride :		50 mg
Dichloran	:	50 mg
Oxgall	:	250 mg
PDA	:	39 G
Distilled water	:	1 lit.

Spread known volume of soil-water suspension over the selective medium and incubate at 25-27^{0}C for 7-10 days.

(ii) Semi-selective medium for *Verticillium chlamydosporium* (De Leis and Kerry, 1991)

Corn meal agar	:	17 g
Rose Bengal	:	75 mg
NaCl	:	17.5 g
Streptomycin sulphate	:	50 mg
Aureomycin	:	50 mg
Chloramphenical	:	50 mg
Triton X-100	:	3 ml
Carbendazin	:	37.5 mg
Thiabendazole	:	35.5 mg
Distilled water	:	1 lit.

Soil is diluted serially by serial dilution technique and a 0.2 ml aliquot of the appropriate dilution is plated on the medium and incubate at 18-25^{0}C and colonies are formed after 5-15 days.

(iii) Trichoderma Selective medium (TSM)

MgSO$_4$. 7H$_2$O	:	0.2 g
KH$_2$ PO$_4$	:	0.9 g
KCl	:	0.14 g
NH$_4$ NO$_3$	:	1.0 g
Anhydrous Glucose	:	3.0 g
Rose Bengal	:	0.15 g
Agar	:	20.0 g
Distilled Water	:	950 ml.

After autoclaving, add 50 ml of antimicrobial agents (chloramphenicol 0.25 g, Quintozone 0.2 g; Captan 0.2 g and Meta laxly 1.6 g). Soil is processed by serial dilution technique and an aliquot of 0.5 ml is added to sterile Petridishes containing selective medium and incubate at 25-27^{0}C and the colonies are observed after 7-10 days.

(iv) Potato dextrose agar (PDA)

Potatoes	:	200 g
Dextrose	:	30 g
Agar	:	20 g
Distilled water	:	1000 ml
pH	:	6.5

Peel potatoes, slice finely,. boil with 500 ml water till soft, strain. Make volume to 1000 ml. After sterilization, PDA is supplemented with 300 µg-1L oxytetracycline.

(v) Special Nutrient Agar (SNA)

KH_2PO_4	:	1.0 g
$MgSO_4. 7H_2O$	:	0.5
KCl	:	0.2 g
KNO_3	:	0.5 g
Glucose	:	0.2 g
Sucrose	:	1.0 g
Distilled water	:	1 lit.

After sterilization, medium is supplemented with 200 mg ml-1L Oxytetracycline.

(vi) Kings 'B' medium

Used for Isolation of antagonistic bacteria viz., *Pseudomonas fluorescens*.

Proteose Peptone	:	20 g
K_2HPO_4	:	2.5 g
Glycerol	:	15 ml
$MgSO_4. 7 H_2O$	:	6 g
Agar	:	15 g
Cycloheximide	:	75 mg
Amplicillin	:	50 mg
Chloramphenical	:	1 lit.
Distilled water	:	1 lit.
pH	:	6.5

(vii) Soil extract Agar medium (Allen, 1957)

Soil Extract	:	1000 ml
Glucose	:	10 g
K_2HPO_4	:	0.5 g
Yeast extract	:	0.5 g
Agar	:	15 g
Distilled water	:	900 ml
pH	:	7.0

Soil extract preparation : one kg of soil is mixed in 1 litre of tap water, autoclave at $121^{0}C$ for 30 minutes and allow to stand overnight. A pinch of $CaCo_3$ is added. The soil is filtered through sterile what man no.1 filter paper and this filtrate can be used for preparation of medium.

(viii) Murashige and Skoog medium (MS medium)

Murashige and Skoog medium is most commonly used for callus induction in tissue culture studies.   The constituents of the medium and the amount of stock solutions to be added to prepare one litre of the medium are given below.

To make one litre of the semi-solid MS medium, the required volume of each stock solution has to be pipetted out into a one litre glass beaker on a magnetic stirrer. Myo-inositol and sucrose are to be added as solids and allowed to dissolve fully adding more water if necessary. Agar to be pre-dissolved by heating in about 250 ml water and then added to the medium component stock solution.

(a)   Preparation of stock MS-IV

FeSO$_4$. 7H$_2$O and Na EDTA are to be separately boiled in 250 ml double distilled water until dissolved and mix later.

(b)   Preparation of hormone stocks

The phytohormones used for callus induction in different composition are auxins (2,4-D, NAA) , gibberellins and cytokinin(Kinetin)

To prepare stock solutions, 50 mg of each hormone has to be dissolved separately, in little quantities of 0.1 N NaOH and then the final volume is made upto 500 ml with double distilled water. The hormone stocks and the vitamins MS-V are to be stored frozen at $-20^{\circ}C$. The inorganic stock solutions can be stored upto one month at $4^{\circ}C$.

Table A.1 Stock solutions of MS basal medium.

Stock No.	Constituents	Stock concn.	Final conc. In the medium	Amount to be taken / lit.
MS-I	KNO$_3$	76.0g/1000ml	1900	25ml
	NH$_4$NO$_3$	66.0g/1000 ml	1650	
	KH$_2$PO$_4$	6.8g/1000ml	170	
	MgSO$_4$.5H$_2$O	14.8g/1000ml	370	

Stock No.	Constituents	Stock concn.	Final conc. In the medium	Amount to be taken / lit.
MS-I	KNO$_3$	76.0g/1000ml	1900	25ml
	NH$_4$NO$_3$	66.0g/1000 ml	1650	
	KH$_2$PO$_4$	6.8g/1000ml	170	
	MgSO$_4$.5H$_2$O	14.8g/1000ml	370	
	KI	20.0 mg /250ml	0.8	25 ml
	Na MOO$_4$.2H$_2$O	6.25mg / 250 ml	0.25	
	CuSO$_4$. 5H$_2$O	0.63 mg/250 ml	0.025	
	CaCl$_2$. 5H$_2$O	0.63 mg/250 ml	0.025	
MS-IV	FeSO$_4$. 7H$_2$O	1.39 g/500 l	27.8	10 ml
	Na$_2$ EDTA	1.865g/500 ml	37.3	
MS-V	Nicotinic acid	20 mg/200 ml	0.5	
	Phyridoxine Hcl	20 mg/200 ml	0.5	5 ml
	Thiamine HCl	4 mg/200 ml	0.1	
	Glycine	80 mg/200 ml	2.0	
	Myo-issoitol	To be added as dry powder		100 mg
	Sucrose	To be added as dry powder		30 g
	Hormones	Stock solution as per requirement		
	Agar	To be added as dry powder		8 g

II Gel Electrophoresis

Reagents and gel preparation for SDS -PAGE

(Laemmli buffer system)

Materials : General lab glassware, vertical slab gel electrophoresis unit, powerpack, syringes (1.0 ml and 20 ml size). Micropipettes, glass tray, light box, parafilm), gel casting stand : marker proteins.

(i) Reagents

Acrylamide / Bis acrylamide (30 : 0.8 g/g) : 100 ml

Acrylamide   :   30.0g

Bis acrylamide   :   0.8 g

Dissolve in distilled water and make upto 100 ml. Filter, if necessary, and store at 4^{0}C in amber colored bottle in dark.

1.5 M Tris – HCl, pH 8.8

Tris Base   :   18.15 g / 100 ml

Dissolve in about 80 ml water adjust pH to 8.8 with 1 N HCl and make upto 100 ml with distilled water, store at 4^0C.

0.5 M Tris-HCl, pH 6.8

Tris base   :   6.0 g / 100 ml

Dissolve in about 60 ml distilled water adjust pH to 6.8 with 1 N HCl and make upto 100 ml with distilled water. Store at 4^0C.

10% SDS

Dissolve 10g SDS in water with gentle stirring and bring to 100 ml with dH_2O.

Sample buffer (store at room temperature) pH 6.8

Distilled water	:	4.0 ml
0.5 M Tris-HCl buffer, pH 6.8	:	1.0 ml
Glycerol	:	0.8 ml
10% (W/V) SDS	:	1.6 ml
2-mercaptoethanol	:	0.4 ml
0.05% (W/V) bromophenol blue	:	0.2 ml

Dissolve the sample in sample buffer or dilute the sample atleast 1:4 (v/v) with sample buffer, and heat at 95^0C for 4 min.

5 × Electrode (tank) buffer, pH 8.3.

		600 ml	150 ml	1000 ml
Tris base	:	9.0 g	450 mg	15 g
Glycine	:	43.2 g	2.16 g	72 g
SDS	:	3.1 g	150 mg	5 g

Dissolve in d H_2O, adjust pH to 8.3, and make upto required volume with d H_2O.

Store at 4^0C. Warm to 37^0C before use if precipitation occurs. Dilute this buffer at 1:4 (v/v) with dH_2O.

(ii)   Resolving gel preparation

Note :

- Bring the temperature of stock solutions stored 4^0C to room temperature. Otherwise air bubbles will be trapped in the gel.

- Accrylamide and bis acrylamide are strong neurotoxins before polymerization. Avoid direct skin contact by wearing disposable gloves.

- Do not mouth pipette the acrylamide solutions. Use pipette pump or rubber propipette.

Table A.2 Resolving (separating) gel composition.

	7.5%	10%	12%	15%
1. Distilled water	14.45 ml	12.05 ml	10.05 ml	7.05 ml
2. 1.5 Tris-HCl, pH 8.8	7.5 ml	7.5 ml	7.5 ml	7.5 ml
3. 10% SDS stock	0.3 ml	0.3 ml	0.3 ml	0.3 ml
4. Acrylamide / Bis stock	6.15 ml	10.0 ml	12.0 ml	15.0ml
5. 10% Ammonium persulfate (APS)	150 ul	150 ul	150 ul	150 ul
6. TEMED	15 ul	15 ul	15 ul	15 ul
	30 ml	30 ml	30 ml	30 ml

Final conc. Of buffer-gel : 0.375 M, pH 8.8

* Prepare just before use by dissolving 100 mg in one ml dH_2O.

Mix the first 4 solutions in a conical flask with side neck, close it with a stopper, and degas at room temperature for about 15 min. and then add solutions 5 and 6, swirl the flask gently to mix and then transfer quickly to gel mould.

Table A.3 Stacking (spacer) gel preparation.

1. Distilled water	6.1 ml	5.68 ml
2. 0.5 M Tris-HCl, pH 6.8	2.5 ml	2.5 ml
3. 10% SDS	100 ul	100 ul
4. Acrylamide/Bis stock (Degas the solution in a side neck conical flask for 15 min at room temp.	1.3 ml	1.66 ml
5. 10% APS (Freshly prepared)	50 ul	50 ul
6. TEMED	10 ul	50 ul
	10 ml	10 ml

Final conc. Of buffer-gel. 0.125 M. pH 6.8

Reagents for staining and destaining

Coomassle Brilliant blue R 250	(0.1%)
Methanol	100 mg
Dissolve, then add acetic acid	50 ml
d H_2O	10 ml
Filter before use	100 ml

Destaining solution :

Methanol	40 ml
Acetic acid	14 ml
d H_2O	146 ml
200 ml	

Gel storage solution : 10% acetic acid.

Sample preparation

- Dissolve purified virus pellets or purified virus structural and non-structural protein pellets directly in sample buffer or dilute the concentrated samples in sample buffer (1:5, v/v)
- Incubate the sample in boiling waterbath for 4 min and cool them in ice-water.
- Use the samples immediately for loading on to the gels or store the samples in small aliquots at -20^0C for further use.
- Process the marker proteins as in step 2.

Slab gel electrophoresis

Procedure

(a)   **Preparation of the separating gel**

- Assemble the vertical slab gel unit in the casting mode. Use 1.5 or 1.0 mm spacers.
- In a 125 ml side arm vacuum flask, mix separating gel solution according to Table A.1 Leave out the APS and TEMED. Add a small magnetic stir bar.
- Stopper the flask and apply a vacuum for five minutes while stirring on the magnetic stirrer.
- Add the APS and TEMED, and gently swirl the flask to mix. Be careful not to generate bubbles.
- Pipette the solution into the sandwich to a level about 4.0 cm from the top.

- Gently apply abut 3.0 ml of water on the side of the slab next to the spacer. The water layers evenly across the entire surface after a minute or two. A very sharp water-gel interface will be visible when the gel is polymerised.
- Tilt the casting stand to pour off the water layer.
- Rinse the surfaces of the gels oncer with distilled water.

(b)   Preparation of the Stacking Gel.

- Pour the liquid from the surface of the
- In a 50 ml side arm vacuum flask, mix stacking gel solution according to Table 2. Leave out the APS and TEMED. Add a magnetic stir ban.
- Deaerate the solution as before.
- Add the APS and TEMED, gently swirl the flask to mix.
- Fill the sandwich with stacking solution.
- Insert a comb into the sandwich. Take care not to trap any bubbles below the teeth of the comb. Oxygen will inhibit polymerization and will cause a local distortion in 10 gel surface at the bottom of the wells.
- Allow the gel to set for at least 30 minutes.
- Put the tube containing the sample in a boiling water bath for 4 minutes.
- Remove the sample and put it on ice until ready to use. This treated sample can be put in the freezer for future runs.

(c)   Loading and running the gel :

- Slowly remove the combs from the gel. Be careful to pull the comb straight up to avoid disturbing the well dividers.
- Fill each well with tank buffer.
- Using a Hamilton syringe, underlayer the sample in each well.
- Fill the lower buffer chamber with tank buffer until the sandwiches are immersed in buffer. If bubbles get trapped under the ends of the sandwiches, coax then away with a pipette.
- Fill in the upper buffer chamber with tank buffer. Take care not to pour buffer into the sample wells because it will wash the sample out.

- Connect the unit to the power supply. The cathode should be connected to the upper buffer chamber.
- Turn the power supply on and adjust the voltage.
- When the dye reaches 1 cm above the bottom, turn the power supply off and disconnect the power cables. Staining and destaining of the gel.
- Disassemble the sandwich and put the gels into stain.
- Gently shake the gels for 1-2 hour.
- Remove the gels and put them in destaining solution until the gel is destained properly.
- Transfer the gels to the gel storage solution.

Precautions

- Acrylamide is carcinogenic and hence gloves must be used for the entire process.
- Use always glass distilled water to prepare the solutions.
- Use good quality chemicals especially acrylamide and bis-acrylamide.

Results : Observations and evaluation.

- Place the slab gels over a light box or place them in a glass tray (containing gel storage solution) kept over a light box.
- Take a photograph of the gels. Draw sketches representing bank pattern in each lane. Measure the distance (from the top (origin) of separating gel to the mid point of each band) migration for each bank and dye front.
- Molecular weight determination.

Example : Virus coat protein

1. Distance migration for each protein bank and tracking dye

Sample	Distance migration (cm)	Relative mobility (Rf)	Mol.wt.
Virus protein			
Marker protein			
Tracking dye			

Calculate the relative mobility (Rf) of a protein or isozyme

$$Rf = \frac{\text{Distance of protein/Isozyme migration}}{\text{Distance of tracking dye migration}}$$

Plot the Rf values (abscissa) against the known molecular weights of marker proteins (ordinate) on semi-logarithmic paper. Estimate the molecular weight of unknown protein from calibration curve.

Plot the distance migration values (abscissa) against he known molecular weight of marker proteins (ordinate) on semi-logarithmic paper. Estimate the molecular weight of unknown protein from calibration curve.

III   Double Antibody Sandwich – Enzyme linked Immunosorbent Assay (DAS-ELISA)

Materials : Purified immunoglobulins (lgG) from virus specific antisera, Purified lgG (same as above) conjugated with enzyme (penicillinase or alkaline phosphatase), Carbonate buffer, coating buffer pH 9.6,PBS-Tween, Antigen buffer, PBS-Tween containing 2% polyvinyl pyrrolidone (40,000 M.W), Penicillin-bromothymol blue substrate, N-para nitrophenyl phosphate substrate.

Procedure

- Coat ELISA plate with antigen specific immunoglobulin @ 100 µl per well.
- Incubate the plate at 37^0C for 3h.
- Take 100 mg of virus infected leaf samples. Grind with five volumes of coating buffer containing two per cent polyvinyl pyrollidine. Centrifuge at 5000 rpm for ten minutes and collect the supernatant.
- Add the supernatant (100 µl) into the plate. Incubate at 37^0C for 3h.
- Wash the microtitre plate three times with the help of PBS-T with five minutes interval.
- Add 100µl of enzyme labeled antibody conjugate to each well and keep the plate at 4^0C for overnight.
- Wash the wells thoroughly with PBS-T for three times with 5 minutes interval.
- Add the substrate, N para nitrophenyl phosphate @ 0.25 mg/ml of diethanolamine buffer pH 9.8 to the wells. Stop the reaction after 30 minutes by adding 3M NaOH at 50µl/well.
- Keep the plate for 30 minutes at room temperature for color development.

- Measure the development of yellow color at 405 nm in ELISA reader.
- Maintain two controls, one with healthy sap and another in the buffer.

IV   Dot Immuno Binding Assay (DIBA)

Materials : Nitrocellulose membranes, Antisera, Micropipettes, Glass dish (Approx. 10x12cm), Shaker, Blunt-tipped forceps.

Solutions Required

(a)   Coating buffer

Na_2Co_3	:	1.59 g
$NaHCO_3$	:	2.93 g

Dissolve in about 900 ml distilled water, adjust pH to 9.6; make up the volume to 1L.

(b)   Tris-buffered saline (TBS)

Tris (0.02M)	:	4.84 g
NaCl (0.15 M)	:	58.48 g

Dissolve in about 1.9 L distilled water, adjust pH to 7.5 and make up the volume to 2 L.

(c)   TBS-Tween

TBS	:	1L
Tween	:	0.5 ml

(d)   Blocking solution :

TBS	:	100 ml
Non-fat dried milk powder :	5 gms.

(e)   Antibody buffer :

TBS	:	100 ml
Non-fat dried milk powder :	5g

(f)   Alkaline phosphatase – labelled goat anti rabbit lgG.

(g)   Substrate for ALP conjugate : BCIP-NBT

5-Bromo, 3-Chloro, 2-Indolyl phosphate –Nitro blue tetrazolium)

Procedure

- Prepare appropriate dilutions of antigen in carbonate coating buffer. Comparable extracts from healthy tissue should be used as controls.

- With a micropipette, apply 5 or 10µl of antigen dilutions onto the membrane.
- Air dry the membrane for at least 15 min.
- Transfer to a lid of a plastic Petri dish and add blocking solution so that the membrane is fully immersed.
- Shake membrane in blocking solution at room temperature for 1 h.
- Weigh comparable healthy plant tissues. Grind tissue at a 1:20 dilution in antibody buffer. Filter through muslin cloth. Use this to prepare an appropriate dilutions of virus specific antiserum. After allowing the diluted antiserum to remain at 37^0C for 45 min. it is ready for use.
- Remove the membrane from the blocking solution and transfer it to diluted antiserum shake at RT for 1 h.
- Pour of the antibody solution. Wash the membrane in TBS – Tween + milk powder thrice, shake for 5 min, at each wash.
- Add diluted goat anti-rabbit lgG labelled with alkaline phosphate Conjugate is diluted @ 1:500 in antibody buffer.
- Shake at RT for 1 h.
- Pour of conjugate solution, wash the membrane as in step (1).
- Add substrate (BCIP-NBT) and shake till colour develops.
- Pour of the stain and wash the membrane in distilled water.
- In case of positive reaction, blue colour will appear as a dot.

Glossary

Abnormality : a deviation from the normal, a malformation or teratology, state of disease.

Abrasive : fine particles of a material such as charcoal, carborundum or diatomaceous earth ('celite') incorporated in the inoculum or dusted on to leaves before inoculation to facilitate the mechanical transmission of a plant virus

Absorption : movement of pesticide (light, water, fluids, etc.,) from a surface into the body. The process by which a chemical is sucked or taken into plants or animals.

Accession : a new member to a plant collection. In the context of plant pathology and plant breeding, accessions are often tested for disease resistance.

Accidental host : an uncommon or rare host to a pathogenic microorganism

Acervulus : (pl.acervuli) (La. Acervus = heap; ul-dim.suf.) an asexual fruiting body.

Acid-fast : resist decoloration by mineral acid.

Acidophile : any organism that grows best under acidic conditions (low pH).

Acquired immunity : immunity which develops in response to foreign antigens in the body, involving either the production of antibodies or activated T cells which react specifically with the foreign antigen

Acquired resistance : (induced resistance or acquired immunity) a resistance response developed by a normally susceptible host following a pre-disposing treatment, such as inoculation with a virus, fungus, bacterium or treatment with certain chemicals. This resistance is not inherited

Acquisition : the entry of virus into, or attachment to, a vector

Acquisition feeding time (or acquisition access time): the feeding time during which a vector feeds on an infected plant to acquire a virus for subsequent transmission (e.g., to become viruliferous)

Acquisition threshold period : the minimum feeding time required for a vector to become viruliferous.

Acrogenous : (Gr. acron=tip; genes-born) developed at the tip.

Acropetal : (Gr. acron-tip; petere= to seek) : produced successively towards the tip, the apical member of a chain being the youngest.

Actinomycete : a group of microorganisms similar to bacteria that produce long filaments. Group of microorganisms apparently intermediate between bacteria and fungi, and classified as either. Characterized by fine hyphae, usually less than 1.0 μm in diameter, that readily break into fragments resembling bacterial cells. Gram-positive, irregularly staining filamentous bacteria with true branching that do not form spores and are non-motile

Activator : a substance that accelerates the effect or increase the total effect of a pesticide.

Active immunity : an immune state achieved by production of antibodies in the host.

Active ingredient (a.i.) : toxic component of a formulated pesticide

Active resistance : resistance resulting from host reactions occurring in response to the presence of the pathogen or its metabolites

Acute : pertaining to symptoms that develop suddenly

Adaption : non-inherited change in an organism towards increased fitness. Also commonly, but incorrectly, used to indicate an inherited change towards greater fitness

Adenosine : a compound one molecule each of adenine and D-ribose and three molecules of phosphoric acid; it plays an important role in energy transformation in metabolism.

Adenosine diphosphate (ADP) : a compound which upon phosphorylation (addition of P and energy) forms high energy bonds like ATP.

Adhesorium : an organ formed from a resting zoospore of *Plasmodiophora* for attachment to and penetration of the host.

Adsorption : condensation of adhesion of gases and liquids on the surface of the solids.

Aeciospore : dikaryotic spore of a rust fungus produced in an aecium; in heteroecious rusts, a spore stage that infects the alternate host

Aerial mycelium : mycelium which grows above the substrate frequenting the air.

Aetiology : the science of the investigation of the cause

Aflaroot : a disease of groundnut caused by the fungus *Aspergillus flavus*

Aflatoxin : chemical byproduct from *Aspergillus flavus* and *A. parasiticus* harmful to humans and other animals Toxin produced by the fungal genus *Aspergillus* which is carcinogenic. They are the derivatives of coumarine. The different aflatoxins are aflatoxin B1, B2, G1, G2 and M2

Agar : a polysaccharide obtained from certain seaweeds (*Gelidium* sp., *Gigartina* sp and *Pterocladia* sp). Agar forms a gel with water, and is used at a concentration of 1.5-2.0% to solidify media used for culturing micro-organisms

Agglutination : a serological test in which viruses or bacteria suspended in a liquid (usually latex particles) clumps together whenever the suspension is treated with antiserum containing antibodies specific against these viruses or bacteria is latex agglutination

Agglutinin : an antibody capable of causing clumping or agglutination of bacteria or other cells.

Aggressiveness : (La. Aggressus from aggredi = to attack): capacity of a parasite to invade and colonize its host plant; often used specifically as a measure of pathogenicity; virulence

Agrobacterium : a type of soil-inhabiting bacteria that is capable of introducing DNA from plasmids in the bacteria into the genome of plant cells. Often used in genetic transformation of plants.

Agrocin 84 : a bacteriocin from the non-pathogenic strain of 84 of *Agrobacterium tumefaciens* used in the control of crown gall.

Allele : one of several alternate forms (DNA sequences) that resides at the same locus on the chromosome and controls the same phenotype.

Alternate host : one of two kinds of plant on which a parasitic fungus (e.g., rust) must develop to complete its life cycle

Alternation of generations : a reproductive cycle in which a haploid phase alternates with a diploid phase.

Alternative host : a plant other than the main host that a parasitic can colonize. alternative hosts are not required for completion of the development cycle of the parasite.

Ambient temperature : the temperature of surroundings averaged over 24 hours.

Amphigynous : having an antheridium through which the oogonium grows, as in many *Phytophthora* species.

Amphitrichous : tuft of flagella arranged at both the end.

Anaerobic : living in the absence of oxygen.

Anamorph : (adj. anamorphic; syn. Imperfect state) the asexual form in the life cycle of a fungus, when asexual spores (such as conidia) or no spores are produced.

Anastomoisis : (pl. anastomoses) fusion between branches of the same or different structures (e.g., hyphae) to make a network.

Antagonist : an organism or substance that limits or counteracts the action of another

Antheridium : (pl.antheridia) male sexual organ (male gametangium) found in some fungi.

Anthracnose : (Gr. Anthrax = coal; nosos = disease): plant disease caused by melanconiales, characterized by producing limited lesions, often sunken in fleshy tissues.

Antibiotic (G. anti = against; biotikos-concerning life) : a substance produced by a microorganism capable of inhibiting the growth of the microorganisms in low concentrations.Examples: Penicillin, Streptomycin, Tetracycline, Blasticidin and etc.

Antibody : a specific protein formed in the blood of warm-blooded animals in response to the presence of an antigen Any substance or body in tissues or fluids of an organism, as blood or serum, which act in antagonism to specific foreign bodies.

Anticodon : a sequence of three nucleotides (in a RNA) complementary to a codon triplet in mRNA.

Antigen : (Gr. Anti = against; gen, genos = birth, growth) : a substance that introduces production of antibody when introduced into the body of the living warm-blooded animal. Any foreign substance that elicits an immune response (e.g. production of specific antibody molecules) when introduced the tissues of susceptible animal and which capable of combining with the specific antibody molecules produced. Antigens are usually high molecular weight and are commonly either proteins or polysaccharides.

Antigen valency : the number of antibody binding sites on the antigen.

Antiserum titre : the greatest dilution of an antiserum which will give a detectable reaction with an antigen. Usually detected by a series of two-fold dilutions 1:2, 1:4, 1:8…. 1:256 and usually expressed as the reciprocal, 2, 4, 8,… 256 etc.

Anti-sporulent : chemical which prevent or reduce sporulation of a fungus but does not kill the fungus.

Antitoxin : an antibody capable of uniting with and neutralizing a specific toxin.

Aplanospore : a naked, amoeboid or non-amoeboid motile cells; a sporangiospore.

Aplerotic : oospores of Pythiaceae-not filling the oogonium.

Appressorium : Pl. appressoria (La. aprimere = to press against) : a swelling on a germ tube or hyphae attaching it to a host tissue in an early stage of infection.

Arbuscule : (adj. arbuscular) branched haustorial structure of certain endomycorrhizal fungi that forms within living cells of the root.

Archaebacteria : most ancient life-forms that persists today.

Ascocarp : (Gr. askos = sac; karpos = fruit) (syn. Ascoma): sexual fruiting body of an ascomycetous fungus that produces asci and ascospores;

Ascogenous : pertaining to ascus-producing hyphae.

Ascogonium : (pl. ascogonia) : a specialized cell that gives rise to the hyphae that produce asci.

Ascomycetes : (adj. ascomycetous) a group of fungi, also called the Ascomycotina, that produces sexual spores (ascospores) within a saclike structure called an ascus

Ascospore : (Gr. askos = sac; sporos=seed, spores) sexual spore produced in an ascus (perfect stage); a meiospore borne in an ascus.

Ascostroma : (Gr. askos=sac; stroma = mattress, bet) (Pl. ascostromata) a fruiting body containing bitunicate (doublewalled) asci in locules (Cavities); usually dark with multiple locules, but sometimes single

Ascus : Pl. asci (Gr. askos=sac) a sac-like structure containing usually eight ascospores, characteristics of ascomycetes.

Aseptate : (syn. non-septate, coenocytic) lacking cross walls.

Attenuated : (La. attenuatus=made thin) reduced capacity of a pathogen to cause diseases, reduced virulence.

Autoecious : (Gr. Autos=self; oikos=house) (of parasitic fungi) completing the life cycle on one kind of host (or) in reference to rust fungi, producing all spore forms on one species of host plant

Autoroph : an organism that synthesizes its nutritive substances from inorganic molecules; e.g. plants capable of photosynthesis

Avirulent : (syn. nonpathogenic) unable to cause disease

Axenic : (Gr. A = not; senos = stranger): culture in the absence of living bacteria or other organisms; pure culture or without another organism being present

Bacillus : a rod-shaped bacterium

Bacteriocin : a protein antibiotic, one or more types of which can be produced and excreted by certain strains of bacteria

Bacteriophage : (Gr. bacterion=staff; phagos = phagous, one that eats) : any virus capable of lysing bacteria through multiplication within the bacterial cell.

Bacteriostat : a chemical or physical agent that inhibits the growth of, but is not lethal to bacterial.

Bacterioid : an enlarged branched bacterium surrounded by a membrane that is of host origin, especially the irregular –shaped bodies of the bacterial genus *Rhizobium*, carry out nitrogen fixation within root nodules on the roots of leguminous plants

Bacterium (pl. bacteria) : a major class of microscopic (0.5-2.0 micron) motile or non-motile, unicellular organisms, the smallest living things able to reproduce themselves, achieved mainly by each cell dividing into two. The genetic code is carried in a tangled coil of DNA known as the bacterial chromosome

Ballistospore : basidiospore that is forcibly shot away from the basidium

Basidiocarp : (Gr. basidion = a small bae; karpas = fruit) (syn. Basidioma) (pl.basidiomata) : fruiting body of the Basidiomycete fungi in or on which basidia are produced

Basidiomycetes (adj. Basidiomycetous; syn. Basidiomycotina) : a class of fungi characterized by exogenous production of spores (basidiospores) on club shaped organs of meiosis (basidia)

Bergey's manual : (of Determinative Bacteriology) an International reference book which classifies and describes bacteria.

Binary fission : a type of asexual reproduction in which two cells, usually of similar size and shape, are formed by the growth and division of one cell

Binomial nomenclature : the method of scientifically naming plants and animals in descriptive Latin terms. The first term identifies the genus, the second the species to which an organism belongs. The first letter of the generic name is capitalized and both names are italicized. The name (often abbreviated) of the author responsible for naming the organism may follow the Latin binomial. When another author transfers species to another genus, the name of the first author is placed in parentheses and the name of the second author follows. Thus, the scientific name of the fungus that causes brown rot of peach is written *Monilinia fructicola* (Wint.) Honey

Bioassay : (Gr. bio = life through; Fr. assay (from La. Exagium = act of weighing) : Determination of chemical effects of tests on living organisms. Also a term applied to a method for determining insecticidal residues employing previously establish dosage mortality figures for a given compound on a suitable test organism. Although extremely sensitive, this method does not distinguish and identify residues of different classes of insecticides

Biocide : a compound toxic to all forms of life

Biodegradation : the process whereby a compound is decomposed by natural biological activity.

Biological control (syn. Biocontrol) : The regulation of plant and animal numbers by natural enemies. It is the aim of biological control to manipulate natural enemies (parasitoids, predators, pathogens in an attempt to reduce the pest numbers and keep them at much reduced levels. The manipulation can involve the introduction of natural enemies into a region where they previously did exist to counter accidentally introduced pests of crops (classical biological control). Also manipulation can involve the use of indigenous natural enemies to augment existing populations or to alter the environment to improve conditions for enhanced natural enemy activity. The use of one biological agent to control another generally by predation or parasitism (Spedding, 1975)

Biological control agent : any biological agent that adversely affects pest species

Bioluminescens : emission of light by living organisms.

Biotechnology : any technique that uses living organisms, or substances from those organisms to make or modify a product, to improve plants or animals, or to develop microorganisms for specific uses (OTA, 1989). Traditional biotechnology covers well established and widely used technologies based on the commercial use of living organisms e.g. biotechnologies used in brewing, food fermentation, conventional animal vaccine production. Modern biotechnology encompasses technologies based on the use of : recombinant DNA technology; monoclonal antibodies and new cell and tissue culture techniques (Persley,1992)

Biotic : (Gr. Biotikos=life) : relating to life, as disease caused by living organisms.

Biotic agent : a living organism that is the cause of damage

Biotic factors : the influence of plants or animals on the environment (opposite of abiotic factors)

Biotroph (syn. Obligate parasite) : obligate parasite i.e., an organism which is entirely dependent upon another living organism as a source of nutrients.

Biotrophic fungi : defined as deriving their energy from living cells of other organisms and most commonly, their host are vascular plants. The resulting plant-fungal association ranging from being mutualistic (e.g., Mycorrhizal roots) to clearly parasitic (Health and Skalamera, 1997).

Biotype : (Gr. bios = life; typas = impression, model) : a subdivision of a species, subspecies, or race based on same identifiable physiological trait such as a specific virulence pattern

Budding : a form of asexual reproduction typical of yeast, in which a new cell is formed as an outgrowth from the parent cell.

Callus : tissue overgrowth around a wound or canker. A hard protuberance; the new tissue produced at the base of a cutting or when a part is severed or injured. Superficial unspecialized tissue produced by plants in response to wounding. Parenchymatous tissue of cambial origin that forms in response to wounding. A mass of thin-walled cells, usually developed as the result of wounding or infection

Callus culture : a mass of undifferentiated cells originating from any type of explant. In a callus, usually developed on nutrient agar, the cells are generated in an unorganized clump, analogous to a pile of loose bricks. In a plant the cells are differentiated and organized systematically to form shoots, roots and other organs

Capsid : The protein coat or shell of a virus particle; the capsid is a surface crystal built of structure units. The protein coat of viruses forming the closed shell or tube that contains the nucleic acid and consisting of protein subunits of capsomeres

Capsomere : a small protein molecule that is the structural and chemical unit of the protein coat (capsid) of a virus

Capsule : a thick slime layer of mucopolysaccharides surrounding the granulosis virus rod. Gummy materials secreted in a compact layer outside the cell wall. A compact layer of polysaccharide exterior to the cell wall in some bacteria

Carrier : a material serving as diluent and vehicle for the active ingredient (e.g., of fungicide), usually in a dust.

Casual agent : organism or agent that incites and governs disease or injury

cDNA : DNA synthesized from RNA using reverse transcriptase enzyme.

Cellulase : (La. cellula = little cell; ase = from ending of diastase, derived from Gr. Disatasis, separation) any of the enzyme that accelerates the hydrolysis of cellulose.

Cellulolytic : (La cellula = little cell; Gr. lysis = a loosing; ic adj.suf.) capable of decomposing cellulose.

cfu : (abbr. for colony forming unit) the number of colonies formed per unit of volumes or weight of a cell or spore suspension

Chemostat : a continuous culture device controlled by the concentration of limiting nutrient

Chemotherapy : the treatment of a plant or animal with chemical to destroy or inactivate a pathogen or parasite without seriously affecting the host

Chimaera : (=chimera) plant or organ consisting of two or more genetically different tissues

Chlamydospore : (La. chlamyd-from Gr.chlamyx = a short mantle; sporos = spores, seed) thick-walled or double-walled asexual resting spore formed from hyphal cells (terminal or intercalary) or by transformation of conidial cells that can function as an overwintering state.

Circulative transmission : (syn. persistent transmission) virus transmission characterized by a long period of acquisition of the virus by a vector (typically an insect), a latent period of several hours before the vector is able to transmit the virus, and retention of the virus by the vector for a long period, usually several days; the virus circulates in the body of the vector

Circulative virus : a virus that is acquired by their vectors through their mouth parts, which then passes through the gut wall of the vector into the haemolymph and eventually contaminates the mouthparts via the saliva

Clamp connection : a bridge or buckle-hyphal protrusion in basidiomycetous fungi, formed at cell division and connecting the newly divided cells

Clone : (Gr. Klon = twig) : a population of cells or organisms of identical genotype; population of recombinant DNA molecules all carrying the same inserted sequence; the vegetative propagation of an organism to produce a population of identical individuals; the use of in vitro recombination techniques to insert a particular DNA sequence into a vector

Codon : a triplet of nucleotides in a DNA or RNA molecule that codes for one of the 20 amino acids in proteins, or for a signal to start or stop protein production. Each gene that code for protein is a series of codons that gives the instructions for building that protein.

Complementary : the opposite or " mirror" image of a DNA sequence. A complementary DNA sequence has an "A" for every "T" and every "T and C for every G". Two complementary strands of single stranded DNA will join to form a double stranded molecules.

Coat protein : the protective layer of protein surrounding the nucleic acid core of virus; the protein molecules which make up this layer

Coenocytic : having multiple nucleic embedded in cytoplasm without cross walls; non-septate

Colonisation : the introduction and establishment of a species (usually beneficial insects) in a new area or the spread of a pathogen in host tissue away from the initial site of infection.

Colony (La. colonis = a settlement) : a group of individuals of the same species living in close association; in fungi, the term usually refers to many hypha growing out of a single point and forming a round or globose thallus.

Compatibility : capable of co-existing in harmony. When two or more chemical insecticides or diluents are mixed together they must not react in such a way that the chemical composition is changed (gelling or precipitation) so as to render the mixture harmful to plants, less toxic to pests, more corrosive to equipment or in some cases, completely useless

Compatible : of a relation between host and pathogen in which disease can develop. When host and pathogen are incompatible disease does not develop, commonly because of hypersensivity of the host.

Competition : occurs when two or more organisms, or population, interfere with or inhibit one another through striving against each other to secure some resource that is in limited supply e.g. the active demand by weeds for nutrients, moisture, light and other essential growth factors adversely affecting crop yields. Occurs when two or more organisms, or populations utilize common resources that are in short supply (exploitation); or if the resources are not in short supply, competition occurs when the organisms seeking that resource harm one another in the process

Competitive inhibition : the inhibition of the action of an enzyme by a non-substrate molecule occupying the site on the enzyme that would otherwise be occupied by the substrate.

Competitive saprophytic ability (CSA) : the summation of physiological characteristics that make for success in competitive colonization of dead organic substrates.

Complement : collective name for a complex of proteins in blood serum that bind in a complex series of reactions to antibody (either IgM or IgG) when the antibody is itself bound to antigens on the cell surface

Complement fixation : removal of complement can be measured and is a sensitive indicator of occurrence of an antigen-antibody reaction

Composite disease : a disease caused by two or more unrelated viruses;

Conjugation : *(La. conjugare=to units)* : in eukaryotes, the process of sexual reproduction by which haploid gametes fuse to form a diploid zygote: in prokaryotes, transfer of genetic information from one cell to another by cell contact.

Contaminant : a microorganism, virus or substance accidentally introduced into a culture, substrate of medium

Cosmid : plasmid vector deisgned for cloning large fragments of eukayotic DNA. The vector contains cos site (=cohesive end sites) that enable it to be packaged in a phage coat *in vitro*

Cross absorption test : a test of serological affinities of two antigens, e.g. viruses. This involves testing dilutions of an antiserum against the antigen used in its preparation after previously incubating the antiserum with an excess of dissimilar antigen

Cross protection : increasing the resistance of a host against pathogen specific to it by pre inoculation of a closely related organisms, e.g. cirus inoculated with mild strains of tristeza virus protects the plants from severe strain.

Cross resistance : resistant strains are frequently cross-resistant to structurally related chemicals or to chemicals with identical mode of action.

Crozier : the hook of an ascogenous hypha before ascus development; the curved apical portion of a blighted stem

Damping off : the rot of seedlings near soil level or prior to emergence (pre-emergence) or after emergence (post-emergence). Disease or necrotic symptom of disease in seedlings in which the seedling is decayed near the soil line and the seedling topples. Damping-off pathogens may also prevent seed germination and kill the sprout before it emerges from the soil

Darkfield microscope : a type of microscopic examination in which the microscopic field is dark and any object, such as organism are brightly illuminated.

Decoration : in this technique the virus particles are attached to the electron microscope grid and then antiserum is added. Homologous antibodies react with the particles to coat or decorate them.

Delphi technique : a forecasting technique carried out in the absence of a relevant database whereby individual experts anonymously give an opinion, and their reasons for it, and the opinions are circulated and amended until a consensus is reached

Demicyclic : a rust fungus that lacks the uredinioscopre (repeating) state (e.g., many species of *Gymnosporangium*)

Dendogram : a tree diagram used as a graphical display to show the similarity of different entities

De-novo : to make a new

Density-gradient centrifugation : a centrifugation procedure in which partially purified virus is further clarified by movement through a gradient. Contamination components may be separated from the virus particles by velocity centrifugation usually in a low to high sucrose gradient, which separates components, according to their differing sedimentation coefficients, or alternatively, by isopynic centrifugation usually in caesium chloride or caesium sulphates gradients, which separates the components according to their differing buoyant densities

Dependent transmission : transmission of a virus (by aphids) that only occurs when the vector feeds on a source plant that is jointly affected by a second virus. The second virus is referred to as a helper virus, and the virus that is not transmissible on its own is called the dependent virus

Detoxification : the inactivation of destruction of a toxin by alteration, binding or breakdown of the toxic molecule. (1) A process, or processes, of metabolism which renders a toxic molecule less toxic by removal, alteration or masking of active functional groups. (2) To treat patients suffering from poisoning in such a way as to reduce the probability and/or severity of harmful effects

Detriment : a measure of the expected harm or loss associated with an adverse event, usually in a manner chosen to facilitate meaningful addition over different events

Deuteromycetes : (syn. Fungi Imperfecti) group of fungi without a sexual stage; the asexual stage; the asexual stage of Ascomycetes and Basidiomycetes

Diagnosis, diagnostic : (Gr. Diagnosis, diagnostikos) : distinctive description, specially the first one, of a taxonomic group determination of pathogen or a disease.

Diapause : a period of spontaneous dormancy, independent of environmental conditions, interrupting development activity in an embryo, larva or pupa

Dictyochlamydospore : a multicelled chlamydospore where the outer wall is separated from the walls of the component cells which are rather easily separated from each other.

Die-back : necrotic symptom of disease in which death of shoot tissues begins at the tip and progresses back towards the main stem

Differential centrifugation : cycles of low and high speed clarification and sedimentation used in the purification of a virus

Differential host (syn. Differential cultivar) (La. Differe = from dis., aparat: ferre = to carry): indicator host, a test plant which by positive or negative response to inoculation of a strain of a pathogen differentiates this from other strains of this pathogen.

Differential medium : a culture medium which can be used to distinguish between different microorganisms

Differential stain : a staining technique which stains different organisms or tissues in different ways so that they may be distinguished e.g. the Gram stain

Differential variety : a variety which gives reactions which distinguish between race specific isolates of a pathogen

Dilution end-point : the lowest dilution in a serial dilution of a virus preparation, that will infect a mechanically inoculated plant

Disease cycle : succession of events in the life cycle of a pathogen and its relationships with its host and the environment that contribute to the process of disease

Disinfectant : an agent that frees from infection by destroying parasites established in plants. This term is best avoided in plant to prevent confusion with its popular use relating to destruction of bacteria etc. on objects or animals, not necessarily in situations where infection is present

Disinfest : to kill pathogens that have not yet initiated disease, or other contaminating microorganisms, that occur in or on inanimate objects such as soil or tools, or that occur on the surface of plant parts such as seed.

Disorder : it refers disturbances of physiological process, and disorder underlies any type of disease.

Dissemination : (La. Disseminare = from disease: seminare = to sow, from semen, seed) (syn.dispersal): Spread of infectious material (inoculum) from diseased plants to healthy plants

DNA chip : a high density array of short DNA molecules bound to a solid surface for use in probing a biological sample to determine gene expression, marker pattern or nucleotide sequence of DNA/RNA.

DNA Probe : a single-stranded DNA molecule used in laboratory experiments to detect the presence of a complementary sequence among a mixture of other single-stranded DNA molecules

DNA profile : the distinctive pattern of DNA restriction fragments or PCR products that can be used to identify, with great certainty, any person, biological sample from a person, or organism from the environment.

Dodder : parasitic seed plant (*Cuscuta* spp). without leaves; a yellow filamentous vine

Dolipore secptum : cross wall found in Basidiomycetes and characterized by special swellings and membranes in association with the septal pore

Dwarfing : underdevelopment of a plant or plant organs, which may be caused by disease, inadequate nutrition, or unfavorable environmental conditions.

Economic threshold : the pathogen density at or above which the value of crop losses (in the absence of management efforts) would exceed the cost of management practices.

Ectomycorrhiza : (pl. ectomycorrhizae) symbiotic association between a nonpathogenic or weakly pathogenic fungus and the roots of plants with fungal hyphae between and external to root cells

Ectoparasite : (Gr.ekto = out, outside) : a parasite that lives on the exterior of its host.

Electron microscopy : electron microscope uses electron beams and magnetic fields to produce the image of high resolution and tremendous useful magnification. There are two types of electron microscope 1. transmission electron microscope (TEM) and 2. scanning electron microscope (SEM)

Electrophoresis : a method of separating substances, such as DNA fragments, proteins by using an electric field to make them move through a "gel" at rates that correspond to their electric charge and size

Enation : (La.ex = out; nation = growing) : abnormal outgrowth from an organ, as a stem or leaf, a symptom of certain virus diseases.

Endemic : (Gr.endemos = from en = in: demos = people; i-c = [adj]suf.) : native to a particular place; pertaining to a low and steady level of natural disease occurrence

Endemic disease : a disease permanently established in moderate or severe form in a defined area, commonly a country or part of a country

Endomycorrhiza : (pl. endomycorrhizae; syn. vesicular-arbuscular mycorrhiza) symbiotic association between a nonpathogenic or weakly pathogenic fungus and the roots of plants in which fungal hyphae invade cortical cells of the root

Endonuclease : nucleases are enzymes that break down nucleic acids into strands of DNA. 'Endo' or inside nucleases act at points along the strand and thus break DNA into short pieces. Endonucleases recognize a particular base sequence in DNA and cut the DNA. Some endonucleases cut the DNA at a specific point while others randomly split the DNA sequences. Specific endonucleases are used by genetic engineers who excise strands of DNA coded for a desirable genetic character. Endonucleases are classed as 'Restriction enzymes' since they are employed e.g., by bacteria to restrict infection by viruses (bacteriophages). The bacterial restriction enzyme attacks the DNA of the infecting organisms

Endoparasite : a parasite that lives within the tissues of the host.

Endophyte : plant developing inside another organism; also used for endoparasitic fungi found in grass species

Endospore : spore forming bacteria in response to adverse environmental condition produce spores, which are highly resistant to various environmental stress.

Endotoxins : substances produced by micro-organisms which are not secreted into the surrounding medium but are confined within the microbial cell; they are released after auto-lysis. A toxin not released from the cell; bound to the cell surface or intracellular

Enzyme-linked immunosorbant assay (ELISA) : a serological test in which the sensitivity of the antibody-antigen reaction is increased by attaching an enzyme to one of the two reactants

EPA : (acronym for Environmental Protection Agency) the federal agency responsible for controlling the various aspects of air, water, and so ill pollution, including pesticide regulations

Epidemic : a widespread temporary increase in the incidence of an infectious disease. Although the term is commonly applied to the rise and fall of disease incidence within a single growth season, e.g. potato blight (*Phytophthora infestans*), yellow rust of wheat (*Puccinia striiformis*) or apple scab (*Venturia inaequalis*), neither the definition nor the concept excludes a similar sequence cover a period of several years or seasons.

Epidemiology : the study of the factors affecting the outbreak and spread of infectious diseases

Epidemiology, comparative : studies in which comparisons are made of the influence of host, pathogen and environment on disease development and spread particularly in so far as these affect initial inoculum or disease, the length of the latent period, the length of the infectious period, the number of dispersal units formed and the effectiveness of these units.

Epidemiology, descriptive : studies, usually of a non-quantitative nature, that depict the behaviour of disease in populations

Epidemiology, quantitative : studies of disease behaviour in populations which utilize measurement or quantification of states and rates of disease processes

Epitope : an amino acid or other sequence that effects formation of an antibody

Eradication : control of plant disease by eliminating the pathogen after it is established or by eliminating the plants that carry the pathogen

Ergotism : a disease caused by ingestion of grain contaminated with alkoloids of ergot fungi, often called the holy fire or st. Anthony's fire in the Middle Ages

Eukaryotes : (Gr, eu = well, true) advanced life-forms, having a variety of sub-cellular organelles, and also possess membrane-bound nucleus

Exoenzyme : extra cellular enzyme; an enzyme secreted by a microorganism into the environment.

Extrachromosomal inheritance : (also cytoplasmic or maternal inheritance) inheritance of genes not located in the nucleus, i.e. those in mitochondria and chloroplasts

Externally seed-borne : (of fungi) located outside functional part of the seed; (of disease) the infection is present on the surface of the seeds but may be externally difficult to eradicate by ordinary seed treatment.

Extra chromosomal genetic element : a genetic element, called a plasmid, that is capable of autonomous replication in the cytoplasm of the bacterial cell.

Facultative : capable of changing life-style, e.g. from saprophytic to parasitic or the reverse

Facultative parasite : organism that is normally saprophytic but is capable of being parasitic

Facultative saprophyte : organism that is normally parasitic but is capable of being saprophytic

Fairy rings : (of a fungus) centrifugal growth of some Gasteromycetes and Hymenomycetes through vegetation near the ground.

Fermentation : the biochemical process of converting a raw material (such as glucose, a sugar) into a final product (such as ethanol)

Field fungi : fungi infecting crop plants in the field, as against fungi developing specifically under the conditions of storage.

Flagellum : (pl.flagell are) : a long whip like structure for the motility of the bacteria.

Fluorescence microscopy : microscopy in which cells or their components are stained with a fluorescent dye and thus appear as glowing objects against dark back ground

Fluorescent antibody : immunoglobulin molecule which has been coupled with a fluorescent dye so that it exhibits the property of fluorescence

Focus (pl.foci) : the site of local concentration of infection or infestation from which secondary spread may occur.

Fomites : inanimate objects that carry viable pathogenic organism.

Foot cells : a basal cell supporting the conidiophore in *Aspergillus*; the basal cell of the conidium in *Fusarium*.

Forecasting : (in pathology) prediction of development of disease as a result of rational study of available pertinent data.

Forma specialis (pl. forma speciales) (La.forma; specialis =special) : a group of biotypes classified on the basis of physiologic properties, particularly as established in relation to a defined host range of plant genera, species or (botanical) varieties (or) a subdivision of a species of a parasitic or symbiotic microorganism distinguished primarily by adaptation to a particular host (Bacterial code, Rec.8a (5) bacterial species have more recently been called pathovar) (or) a taxon characterized from a physiological (especially host adaptation) point of view but scarcely or not all form a morphological standpoint

Fractional sterilization : the sterilization of material by heating it to 100^0C (212^0F) on 3 successive days with incubation periods in between.

Fragmentation : a method of asexual reproduction: the segmentation of the thallus into a number of fragments each of which is capable of growing into a new individual.

Free cell formation : the process by which the 8 nucleic, each with some adjacent cytoplasm are cut off by walls in the immature ascus to become ascospores.

Freeze-drying (syn. Lyophilization) : technique used in the preservation of living material, whereby water is removed under vaccum while the tissue remains frozen.

Freund's complete adjuvant : contains pharmaceutical grade white mineral oil such as parafin, an emulcifier such as mannide mono-oleate, and heat killed *Mycobacterium tuberculosis*.

Freund's incomplete adjuvant : a substance containing an emulsifier and mineral oil, which is mixed with a virus before it is injected into the muscles of an animal to produce anti-serum. The adjuvant allows slow release of the virus following injection

Fructification : general term for spore-bearing organs in both macro-and micro fungi

Fruiting body : a complex structure that bears fungal spores, e.g. sporangia. Coremia, sporodochia, aceruvuli, pycnidia, apothecia, perithecia

Functional genomics : the field of study that attempts to determine the function of all genes (and gene products) largely based on knowing the entire DNA sequence of an organism.

Fungi Imperfecti (syn. Deuteromycetes) : form class erected to contain those fungi whose sexual reproduction is unknown

Gel diffusion test : (syn. Ochterlony gel diffusion test) a serological test in which antibody and antigen reactants diffuse towards each other in gel and react to form visible precipitin line. This test is useful for identification of pathogens as well as to distinguish pathovars of bacteria and strains of viruses.

Gelatin : a protein obtained from skin, bones, tendons etc. used in culture media for the determination of a specific proteolytic activity of microorganisms or for the preparation of a peptone.

Gel chromatography : a molecular sieving procedure, by which viruses are separated from different sized molecules when passed through the pores of gel beads such as agarose; used for virus purification.

Genome : the entire hereditary message of an organism. The total genetic composition of the chromosomes in the nucleus of a gamete. The nucleic acid component of a virus

Graft transmission : transmission of a pathogen from one host plant to another through fusion of living tissue from the diseased host with living tissue of a healthy host

Gram stain : procedure used for identification of bacteria in which crystal violet stain, gram's iodine, ethyl alcohol and safranin stain are applied in succession to cells of the bacteria

Growing-on- test : test on plants grown from seed beyond the seedling stage in a green house on an environment-controlled room.

Growth regulator : a chemical substance produced in one part of an organism and transported in minute quantities to induce a growth response in another part eg. auxins, gibberellins and cytokinin

Hapten : a substance that can combine with a specific antibody but which lacks antigenecity i.e. it cannot initiate an immune response unless bound to an antigenic carrier molecule such as bovine serum albumin.

Histopathology : a study of abnormal microscopic changes in the tissue structure of an organism

Homologous antigen : an antigen reacting with the antibody it had induced

Hybridoma : an antibody secreting cell line which is immortal. A hybrid cell resulting from the fusion of a tumour (cancer) cell and a normal cell such as a lymphocyte from the spleen. The fused cells can be cloned and being derived from a simple spleen cell, will secrete a pure antibody

Immune response : the development of specifically altered reactivity following exposure to an antigen. This may take several forms eg. Antibody production, cell mediated immunity immunological tolerance. Substances capable of inducing the immune response are called immunogens

Immune system : the system in mammals which responds specifically to the presence of foreign antigens to aid in eliminating them from the body; provides acquired immunity

Immunization : the process of increasing the resistance of the host. In plants, complete resistance to disease

Immunity acquired : a non-inherited immunity acquired by some predisposing treatment, modifying the chemistry of plant. The protection afforded to some species by pre-inoculation with non-virulent microorganisms or viruses prior to inoculation with virulent ones is not immunity if any degree of infection by either strain is involved

Immunity : a type of resistance to attack; usually considered an acquired state in which an organism is capable of resisting a pest and thus preventing the development of a disease or damage. In animals it induces the production of antibodies. In plants, the ability to remain free from disease because of inherent structural or functional properties.

Immunoassay : a technique which detects proteins by using an antibody specific to that protein

Immunodiffusion : a serological procedure in which the antigen- antibody reaction is carried out allowing the reactants to diffuse in gel

Immunoglobulin(IgG) : a member of the largest family of proteins produced by animals.

Immunology : the study of acquired immunity in animals and man against infectious disease

Immunosorbent electron microscopy : techniques involving the visualization of antigen antibody reaction in the electron microscope.

In vitro : carried out under laboratory conditions, in test tubes, Petridishes etc.

In vivo : under natural conditions in contrast to artificial, laboratory conditions.

Inclusion body : structure developed within a plant cell as a result of infection by a virus, often useful in identifying the virus.

Incompatibility : failure to obtain fertilization and seed formation after self-pollination, or within or between clones.

Incompatible : sprays which in a tank mix react with each other or the materials of which the equipment is built. Biologically incompatible sprays produce undesirable side effects on the crop when applied together. In plant pathology this term means that a pathogen is unable to form a parasitic relationship with a host plant usually since it has triggered the resistance responses of the host.

Incubation : in seed health testing maintaining the seeds in an environment favourable to development of symptoms or structure of pathogen.

Incubation period : the period of time elapsing between entrance or introduction of micro-organisms in the host and the development of symptoms and signs of an infectious disease.

Indexing : process of determining the presence of disease in a plant by transferring inoculum from the plant to another in which diagnostic symptoms develop. The second plant is termed a 'test' or an 'indicator plant.

Indexing techniques : testing by inoculation of indicator plants to reveal pathogenicity of inoculum obtained from plant material to be tested for presence of pathogens.

Indicator : a plant that reacts to certain viruses or environmental factors with production of specific symptoms and is used for detection and identification of these factors.

Indicator plant : one that reacts viruses or environmental factors with specific symptoms and is used for identification of the viruses or the environmental factor

Indicator, virus : plant used to demonstrate symptoms of specific virus infection

Induction : the process by which an enzyme is synthesized in response to the presence of an external substance, the inducer. Also the activation of an occult pathogen, leading to progressive infection and disease; in particular, the provoked transformation of a provirus into a virulent (cytocidal) virus.

Infect : to invade and establish a pathogenic relationship in a host or to persist in a carrier.

Infection : the act or result of invading and getting established a pathogen in a host or a carrier.

Infection chain or cycle : the never-ending sequence of infection, colonization, sporulation, dispersal and again, infection

Infection court : place in a plant where infection may occur, as flower, fruit, leaf, etc.

Infection cushion : organized mass of hyphae formed on the surface of plant from which numerous infective hyphae develop

Infection peg : a slender structure formed by the deposition of substances such as lignin around a thin fungal hypha penetrating a host cell

Infection thread : the specialized hypha of a pathogenic fungus that invades tissue of the susceptible plant. Also, the primary colonization and inward movement of *Rhizobium* cells within the root hair of a leguminous plant

Infectious : pertaining to disease, capable of spreading from plant to plant

Infective : an organism or part there of capable of establishing a pathogenic relationship with a host i.e. capable of transmitting inoculum

Infectivity assay : a bioassay using mechanical sap-transmission to quantitatively determine the amount of infectious virus

Infest : with insects to occupy and cause damage; referring to soil, contaminate by fungi, eelworms or insects to be present in numbers. To introduce a pathogen into the environment of a host. Infestation does not imply disease and is not be confused with infection

Infestation : presence of a pathogen in material, from which it may infect plants, as in soil or on the surface of seed. of infection, applicable to living, diseased plants or carriers only.

Infiltration : the introduction of liquid under pressure, e.g. into stomata, by spray or by applying a vacuum and then releasing it.

Infra-red spectrophotometry : a widely used technique employed in the identification of organic compounds by passing different wave lengths of the infra – red region of light through solutions in suitable solvents

Ingredient, active : an ingredient in a pesticide or its formulation which is not pesticidally active, i.e. water, emulsifying agent, diluent, carrier etc

Inoculation : applying nodule bacteria on legume seeds just before planting them to transfer inoculum to or into an infection court

Inoculation feeding period : the length of time a vector feeds on the test host in transmission experiments. Synomyous with transfer feeding time

Inoculation threshold period : the minimum feeding period a vector needs on a test plant to transmit a virus

Inoculum density : a measure of the number of propagules of a pathogenic organism

Inoculum potential : The power of a certain amount of inoculum, as defined by testing procedures, to produce disease.

Inoculum : pathogen or its parts, capable of causing infection when transferred to a favourable location

Interference : the interference one virus with the multiplication or the disease-producing capabilities of a second, unrelated virus

Interferon : a naturally occurring anti-viral agent, produced as a result of virus infection or the presence of some nucleic acids that interferes with virus replication

Isoelectric point : the pH at which virus particle has a zero net charge

Isoelectric precipitation : the precipitation from suspension of a virus or a protein when the pH is at the isoelectric point. i.e. when there is no positive or negative surface charge on the virus or protein

Isoenzyme : any one group of enzymes of different structural forms that posses identical catalytic properties.

Isogamy : the condition in which gametes are morphologically similar, as in the members of the subclass Zygomycetes of the class Phycomycetes

Isolate : a culture or subpopulation of a microorganism separated from its parent population and maintained in some sort of controlled circumstance ; to remove from soil or host material and grow in pure culture

Isolation : the procedure by which a pathogen is separated from its host and its culture on a nutrient medium. Also, confinement of a hospital patient to reduce the spread of infectious disease either from. or to other patients

Karyogamy : fusion of two sex nuclei after fusion between the two sexual cells.

Karyotype : the size and number of chromosomes in an organism.

Kilobase : 1000 base pairs of nucleic acid (NA)

Klendusity : escape of infection under natural conditions inspite of susceptibility.

Koch's postulates : the procedure used to prove the pathogenicity of an organism, i.e. its role as the causal agent of a disease

Lag phase : the earliest phase of growth after inoculation, during which the cell number stays constant while cells adjust to the new medium.

Laminar flow : a flow in which the fluid moves smoothly in parallel streamlines; non-turbulent.

Latent : present but not manifested or visible, as a symptomless infection by a pathogen

Latent infection : infection unaccompanied by visible symptoms

Latent period : the time between infection and the appearance of symptoms and / or the produce of new inoculum; the time after a vector has acquired a pathogen and before it can be transmitted

Latent virus : a virus that does not induce symptom development in its host. A virus present in a cell, yet not causing any detectable effect

Latex agglutination test : when homologous antigen is mixed with the sensitized purified gamma globulin, the latex particles aggregates.

Lesion : a usually well-defined abnormal change in structure of an organ due to disease or injury.

Lesion test : a test to determine the ability of an organism to cause a necrotic lesion on plant material, e.g. bean pods, usually from a high inoculum dose. This test is not indicative of ability to infect in the field

Lignification : hardening of tissue through the deposition of lignin in the cell wall

Local lesion : small, restricted lesion, often the characteristic reaction of differential cultivars to specific pathogens, especially in response to mechanical inoculation with a virus

Local lesion assay : a quantitative estimation of the infectivity of preparation of a pathogen, e.g. suspension of bacteria or viruses, from the number of lesions produced in inoculated leaves. Half-leaves are the common experimental unit, and the leaves of an opposite pair differ less than leaves from different nodes; hence the terms half-leaf test and opposite leaf test

Local lesion host : a host which develops local lesions on inoculation with a virus

Local lesion reaction : response of inoculation of leaves by a suspension of bacteria or viruses or some other preparation of a pathogen, on which a quantitative estimation of the infectivity can be recorded.

Lyophilisation : long-term preservation of micro-organisms or tissues by quick freezing at low temperatures and desiccation under high vacuum

Lysogenic bacterium : a bacterium infected with phage, which is integrated with the bacterial genetic material. All progeny are likewise infected and have the capacity to produce phage (by occasional cells undergoing lysis spontaneously or after application of inducing agents, e.g. ultraviolet light). Lysogenic bacteria are not sensitive to the phage they carry

Lysozyme : an enzyme in body fluids, active in host defense, able to kill invading organisms by digesting the cell wall.

Macroconidium : the larger of two kinds of conidia formed by certain fungi

Macrocyclic : a rust fungus that typically exhibits all five stages of the rust life cycle.

Major gene resistance : genetic resistance to disease based on one or a few genes

Mechanical inoculation : a method of experimentally inoculating a plant with a virus through transfer of sap from a virus-infected plant to a healthy plant. Of plant viruses, a method of experimentally transmitting the pathogen from plant to plant; juice form diseased plants is rubbed on test-plant leaves that usually have been dusted with carborundum or some other abrasive material.

Mechanical transmission : spread or introduction of inoculum to an infection court (wounding) by human manipulation, accompanied by physical disruption of host tissues.

Meristem : plant tissue characterized by frequent cell division, producing cells that become differentiated into specialized tissues

Meristem culture : a cell culture developed from a small portion of the meristem tissue of a plant. Either a stem shoot or root meristem can be used

Meristem tip : the meristem dome and first pair of leaf primordial which is the usual unit (0.61-0.65 mm long) excised for aseptic culture, e.g. in attempts to produce virus free plants. Sometimes referred to erroneously as the meristem, or ambiguously as the shoot tip

Metabolic pathway : a series steps in the chemical transformation of organic molecules.

Metal shadowing : a technique used to prepare viruses for electron microscopy, in which the virus particles are exposed to the vapour of a heavy metal such as gold or platinum. Now replaced by the negative contrast staining method

Micro-aerophiles : a species of bacteria which grow best with only a small amount of oxygen levels for growth

Micro-capsule : generally composed of a core material and an outer wall. The outer wall isolates the core material from the environment and protects it from environment degradation and interaction with other materials. The core materials are designed to be released in a controlled fashion

Microclimate : the climate of a usually small site of habitat, as to the climate between and immediately above the plant of an individual crop.

Microconidium : a type of small conidia; some types of microconidia act as spermatia.

Microcyclic : describing a rust fungus that produces only teliospores and basidiosproes

Microflora : vegetation of microorganisms on a restricted area of substrate, as of a plant or in a seed sample after incubation; also that of a geographic region.

Monoclonal : produced from a single genetically pure cell type; often referring to antibody produced from a hybridoma cell made from a single type of B cell

Monoclonal antibody : an extremely pure antibody derived from a single clone of an antibody-producing cell. Invading pathogens, viral or bacterial, carry a large number of different antigens each capable of stimulating the host's immune system to generate a corresponding antibody. A single spleen cell exposed to a specific antigen can be fused with a myeloma cell. The resultant fused cell, called a hybridoma, continually produces an antibody specifically directed against the antigen. It will therefore seek out and identify the specific antigen. Hybridomas can be cloned and cultured to produce quantities of the pure monoclonal antibody. Because of its specificity each monoclonal antibody may be used for diagnostic or therapeutic purpose. Most research and development has employed mouse antibodies grown in the peritoneal cavity of immuno-suppressed mice.

Specific diagnostic tools which allow the rapid detection of individual proteins produced by the cells

Monocyclic : having one disease or life cycle per growing season

Monoecious : refers to plants that have separate male and female flowers on the same plant, synonym of autoecious; also to rusts that have all stages of their life cycle on a single species of plant

Mosaic : disease symptom characterized by non-uniform coloration, with intermingled normal, light green and yellowish patches, usually caused by a virus; often used interchangeably with mottle

Movement protein : for viruses, a protein encoded by the pathogen that is required for spread within an infected plant

mRNA : messenger RNA, a molecule which is synthesized from a DNA template by the enzyme RNA polymerase

Multi-component virus : a virus whose genome is divided into two or more parts each part being separately encapsidated. Hence two or more components are needed to initiate an infection. Note that this is different to a multi-partite genome where components may be enclosed in a single particle

Multi-line : a crop grown from a mixture of seed of almost genetically identical breeding lines which have all agronomic characters in common, but differ in major gene resistance, thereby confronting pests with a mixture of host gene-type

Multiple parasitism : the situation in which more than one parasitoid species occurs simultaneously in or on the body of the host

Multiple regression analysis : a regression analysis in which the mean value of a dependent variable y is assumed to be related to a set of independent variables x1, x2,.....xk by an expression of the form $E(y/x1,x2,......xk) = ß + ß0+ ßx1+ ßx2+ ßxk$

Mutant : an individual having an abnormal structure, property of behavior in which it differs distinctly from the parent organism from which it was derived. This is an inheritable change (a change in the sequence or chemistry of the purine or pyrimidine bases contained in DNA molecules) and may be caused or induced by high energy irradiation or by certain chemical substances

Mutation : a suddenly occurring heritable change in DNA. Such mutations may be brought about by chemical or physical agents e.g. U.V. radiation. An abrupt appearance of a new characteristic in an individual as a result of an accidental change in genes or chromosomes.

Mycoplasma : genus of microorganisms, which have a unit membrane and not a rigid cell wall such as bacteria.

Mycorrhizal : literally, 'fungus root'. A symbiotic association of roots with a fungus which may form a layer outside the root (ectotrophic) or within the outer tissues (endotrophic)

Mycosis : an infection by a parasitic fungus, or a disease caused by the action of mycotoxins

Mycotoxin : any toxin produced by fungi and affecting men and animal.

Myxomycetes : a class of fungi characterized by amoeboid vegetative protoplasts, plasmodia, and by brightly coloured spore-bearing capilitia

Nanometer : one trillionth of a meter, or 10^{-9}; a unit used to measure viruses or individual features of microbial cells

Natural immunity : immunity due to innate genetical character.

Necrophyte : a fungus (or other organisms) colonizing already dead tissues

Necrosis : a localized and rapid destruction of a cell or more often a group of cells and a consequent quick death of those which are in contact with or form part of a living tissue; rot and canker are examples of necrotic symptoms. Cell death (used particularly for death of cells in a focal point in a multi-cellular organism) due to anoxia or local toxic or micro-biological action

Necrotroph : a parasite that typically kills and obtains its energy from dead host cells

Negative contrast staining : a staining procedure used to prepare virus particles for examination in an electron microscope

Negative sense RNA : ribonucleic acid complimentary to the positive or plus sense, not translated into protein

Negative stain : a procedure in which the background is stained whereas the specimen is not

Nematode : non segmented roundworm (animal), parasitic on plants or animals, or free living in soil or water

Non-persistent transmission : a type of insect transmission in which the virus is acquired by the vector after very short acquisition feeding times, and which is transmitted during very short inoculation feeding periods. The vector remains viruliferous for only a short period unless it feeds again on an infected plant

Non-sense mutation : a mutation that changes a normal codon into one which does not code for an amino acid

Non-specific toxin : a metabolic product of an organism which will induce toxic symptoms in a range of different organisms. Most established toxins are non specific both in the type of response and the range of organisms affected, e.g. fusaric acid (n butyl picolinic acid) produced by various *Fusarium* spp. non-virulent; a variant of a pathogenic organism that is incapable of causing severe disease

Nucleic acid hybridization : a joining together of a strand of nucleic acid (DNA or RNA) from one organism with a strand from another by base sequence homology

Nucleocapsids : the structure composed of the capsid with the enclosed viral nucleic acid; some nucleocapsids are naked, others are enclosed in an envelope

Nucleoid : resembling a nucleus. The region in a prokaryotic cell where the DNA is located. Although frequently diffuse, sometimes the DNA is sufficiently contracted so that a defined region can be detected.

Nucleotide : a subunit of a nucleic acid, either DNA or RNA

Obligate parasite : a parasite which is obligated to live in or on its host, as distinct from a facultative parasite, which has the ability to live on or in the host or away from it

Ochratoxin : toxins of *Aspergillus ochraceous, Penicillium viridicatum*; the cause of nephrotoxicosis in sheep, cattle and pigs; also carcinogenic and found in coffee.

Okazaki fragments : DNA strands replicated in small pieces.

Oligogenic : on inheritance of characters determined by few genes.

Oligosaccharide : a compound consisting of relatively few (2-10) mono-saccharide units, which, upon complete acid hydrolysis, yields only simple sugars

Oomycete : a fungus that produces oospores. An order of the phycomycetes

Oosphere : an unfertilized egg, a female gamete

Oospore : a thick-walled resting spore produced by sexual reproduction, as in Oomycetes.

Ooze, bacterial : a mixture of bacteria, secreted mucopolysaccharide etc. so extruded

Operon : a group of genes which function as a collective unit. A cluster of genes whose expression is controlled by a single operator

Opportunistic pathogen : an organism that only causes disease when host resistance is abnormally low

Paragynous : having the antheridium contact the oogonium on the side, as in many *Pythium* sp.

Parasexuality : a process in which plasmogamy, karyogamy and haploidisation takes place in sequence, but not at specified points in the life-cycle of an individual. Of significance in heterokaryotic individuals which derive some of the benefits of sexuality from a parasexual cycle.

Parasite, ecologically obligate : a parasitic organism which can also grow saprophytically, but only if it is not in competition with less-specialized parasites or wholly saprophytic fungi and bacteria

Parasite : an organism which lives in or on another living organism of a different species from which it draws its nourishment.

Pasteurization : the process of heating a liquid or solid food to a controlled temperature to enhance the keeping quality and destroy harmful microorganisms.

Pathogen : a casual agent or factor of disease.

Pathogenesis : production and development of disease

Pathogenicity, horizontal : Robinson (1969) has proposed the use of this term, but defines it only by analogy with horizontal resistance. It can be defined as pathogenicity which is effective against all varieties of the host; or as pathogenicity expressed without differential interactions between races and varieties; cf. vertical pathogenicity. The term is applicable only to a characteristic of a genus or species however, its use is not recommended because the implications of horizontal are too far removed from the biological situation to which it refers

Pathogenicity, vertical : pathogenicity which is effective against some varieties of the host but not against others; pathogenicity expressed in differential interactions between races and varieties.

Pathogenicity : the ability of the pathogen to cause disease.

Pathotoxin : a host-specific toxin.

Pathotype : a parasite population in which all individuals have a given parasitic ability in common i.e. they are distinguished by common characters of pathogenicity, particularly in relation to host range

Peritrichous : having a number of flagellae over the surface of a peritrichous flagellae of bacterium

Persistence : the length of time a virus is retained by a vector after being acquired from an infected plant. Persistence is a crucial feature which has been used to establish three categories with which other characteristics of transmission are associated. Viruses are referred to as non-persistent, semi persistent or persistent according to whether they are retained by feeding by vectors for hours, days or weeks respectively.

Persistent transmission : a type of insect transmission in which the virus is acquired by the vector only after a long acquisition feeding period, and in which there may be a latent period following the acquisition feed, the vector can transmit the virus. The vector remains viruliferous for a long period, often throughout its life span. The virus sometimes multiplies within the vector

Persistent virus : one which is retained for long periods, and sometimes through life, by vectors transferred from infected to healthy

Phage typing : a method of distinguishing between different bacterial species or types within a species by testing reaction to a number of selected phages

Phaseolotoxin : a toxin from *Pseudomonas syringae pv. phaseolicola*

Physiologic race : a biotype and group of biotypes of the same species and variety, morphologically indistinguishable, but which differ in physiological behaviour.

Phytoalexin : a substance in plants that inhibits the development of microorganisms produced in response to chemical or physical injury, or when they are infected by a fungus.

Plasmid : a circular, self-replicating hereditary element that is not part of a chromosome; plasmids are used in recombinant DNA experiments as acceptors and vectors of foreign DNA

Plasmodesma : a strand of cytoplasm passing through a pore in the cell wall, thus usually joining the protoplasm of adjacent cells.

Plasmogamy : a fusion between two sexual cells.

Pleomorphic : having more than one independent from or spore status.

Polyclonal antibody : an antibody preparation that is the product of more than one clone of plasma cells. Such antibodies react with different components of the antigen

Polycyclic : having more than one cycle of infection during a growing season

Polyetic pathogen : one in which the inoculum does not increase during a single growing season but does increase from season to season

Polygenic : on inheritance of characters determined by many genes.

Polygenic resistance : resistance to parasites based on many genes, i.e. a series of genes controlling a quantitative character

Polymerase chain reaction : the reaction leads to the amplification of specific DNA sequences by enormous factor

Positive sense RNA : RNA that can serve directly as messenger RNA

ppm : parts per million (pars in 106 parts) e.g. the number of parts of active ingredient per million parts of the substance in question. They may include residues in soil, water or whole animals. 10 ppm is equivalent to 0.001 %

Precipitin test : a serological test in which the reaction between soluble antigen and antibody results in the formation of a visible precipitate

Predisposition : condition of being susceptible to establishment of a disease.

Pre-emergence : the condition of ungerminated seed and nearly germinated seed below the soil surface; pre-emergence killing.

Primary infection : the first infection following a resting or dormant period of a pathogen.

Primer : a short oligonucleotide that base-pairs to a region of a single-stranded template oligonucleotide. Primers are used to copy adjacent sequences of mRNA and cDNA's

Primer, degenerative : a degenerative primer is actually a mixture of primers, all of similar sequence but with variations at one or more positions.

Probe : a piece of labeled DNA or RNA which is used to locate another piece of nucleic acid by complementary base pairing. The reaction is monitored by autoradiographic or enzymatic detection of the hybridization

Prokaryotes : organisms that lack a clearly defined nuclear membrane.

Propagative virus : a virus which multiplies in its vector

Propagule : any unit in service of propagation, such as spore, sclerotium, mycelial fragment, etc.

Protoplast fusion : any induced or spontaneous union between two or more protoplasts to produce a single, bi-or multi-nucleate cell. In particular, a tissue culture procedure for somatic hybridization that is used in cell manipulation studies.

Psychrophiles : an organism able to grow at low temperatures. An organism that grows best at cold temperatures

Pure culture : an organism growing in the absence of all other organisms

Pure line : a strain in which all members have descended by self-fertilization from a single homozygous individual

Purification : the separation of virus particles in a pure form free from cell components, and their concentration

Pustule : a blister-like fungal spore mass breaking through a plant epidermis. A pimple like, eruptive fruiting structure, such as a uredinium of a rust fungus

Pycnidium : an asexual, hollow fruiting body in which conidia are produced, characteristic of the Sphaeropsidales.

Quarantine : all operations associated with prevention of importation of unwanted organisms into a territory or that exportation from it   e.g the holding of imported material in isolation for a period to ensure freedom from diseases and pests. a) a legal action in which there is a control on the import and export of plants to prevent spread of pathogens and pests. b). holding of imported plants in isolation for a period to ensure their freedom from diseases and plants. Control of import and export of plants to prevent spread of diseases and pests

Race non-specific resistance : resistance to all races of a pathogen

Race-specific resistance : resistance to some races of a pathogen, but not to others

Radial diffusion : an immuno-diffusion serology test in which liquid antigen (or antibody) is placed in a well cut in gel containing the other reactant and allowed to diffuse out into the gel.

Radio-immunoassay : an immunological assay employing radioactive antibody for the detection of certain antigens in serum

Randomized block design : the design used in experimental work in which the treatments are arranged in random order within the replicates blocks to eliminate bias

Recombinant DNA (r-DNA) : a strand of DNA synthesized in the laboratory by splicing together selected parts of DNA strands from different organisms or by adding a selected part to an existing DNA strand

Recombination : process by which genetic elements in two separate genomes are brought together in one unit. The occurrence of progeny with combinations of genes other than those that occurred in the parents, owing to independent assortment or crossing over (after King & Stansfield, 1990). Also, a new strain of a virus that occurs as a result of the breakage and renewal of co-valent links in a nucleic acid chain, so that the nucleic acids are rearranged in the chain

Replication : the process by which a virus particle induces the host cell to reproduce the virus; the process by which a DNA or RNA molecule makes an exact copy of itself; repetition of an experiment or procedure at the same time and place (one of several identical experiments, procedures, or sample)

Resistance, major gene : resistance controlled by genes which have a large observable effect. Not synonymous with oligogenic resistance. The classification of genes into major and minor determinants is arbitrary and depends on subjective judgement of the relative size of the effects observed

Resistance, minor gene : resistance controlled by genes which have small observable effects. Not synonymous with polygenic resistance

Resistance, monogenic : (digenic etc.) resistance controlled by one (two etc.) genes. Resistance controlled by a few genes is termed oligogenic

Resistance, multigenic : synonymous with polygenic but not recommended because of its mixed etymological derivation

Resistance, Oligogenic : resistance controlled by few genes (cf. resistance, monogenic) the phenotypic effects of which may be large or small, i.e. major or minor genes.

Resistance, qualitative : resistance that is expressed in large and discontinuous increments. It is not substantially influenced by environmental factors within normal limits. It is commonly associated with oligogenic inheritance. The term should not be used as an alternative to oligogenic or major gene resistance because it refers to the manifestation of resistance and not to its inheritance. It is not synonymous with vertical resistance. The term should only be used when there is insufficient information to describe the form of resistance more precisely.

Resistance, quantitative : resistance that is expressed as a continuous variate between maximum and minimum levels. It is frequently sensitive to environmental effects, inoculum density etc. Quantitative resistance is generally associated with polygenic inheritance, but the term is not synonymous either with polygenic or minor gene resistance because they refer to different attributes, the expression of resistance and is inheritance, respectively; nor is the term equivalent to horizontal resistance. It is useful only when there is insufficient information to describe the form of resistance more precisely.

Resistance, race non-specific : resistance to all races of a pathogen. The term is essentially synonymous with horizontal resistance and is preferred because it is more obviously descriptive of the condition to which it refers. The form of resistance is commonly, but not necessarily, polygenic and the two terms are not interchangeable.

Resistance, race specific : resistance to some races of the pathogen, but not to others. Usually related to cases of oligogenic resistance where there is a gene-for-gene relationship between host and pathogen. The term is preferable to the synonymous vertical resistance because it is descriptive of the condition to which it refers. It should not be used as a synonym for oligogenic resistance, for the latter is not necessarily race specific; the terms relate to different attributers of resistance.

Resistance, vertical : resistance to some races of a pathogen but not to others (J.E.van der Plank, Plant diseases : epidemics and control, 1963, New York & London, Academic press). Resistance in which significant interactions are detectable in an analysis of variance on a series of pathotypes and a corresponding series of pathodemes (Robinson, 1969) ; cf. horizontal resistance. The use of horizontal and vertical in this context derives from a graphical representation by van der Plank (1963) of the relationships between races of a pathogen and varieties of its host (pathotypes and pathodemes, according to Robinson 91969). Vertical resistance is often expressed as a qualitative modification of disease development, e.g., as a hypersensitive reaction to infection. Vertical resistance is not synonymous with Oligogenic, major gene or active resistance because it refers specifically to the interactions between races and varieties and not to the mode of inheritance or the mechanism of resistance. However, vertical resistance is exactly equivalent to the more familiar and obviously descriptive race specific resistance, and the latter is recommended in preference to the former; cf. resistance, horizontal.

Resistance, general (generalized) : resistance to more than one disease. Sometimes used synonymously with race non-specific resistance but this use is not recommended

Resolving power : the ability of the microscope to distinguish fine details in a microscope specimen.

Restriction enzyme : an enzyme that cuts and effectively excises a piece of DNA. Some restriction enzymes cut the DNA at specific points, others appear to cut at random. Restriction enzymes, of which many hundreds have been identified and isolated, are important tools in the excision and transfer of specific gene sequences from one organism's DNA to another's.

Resurgence : an increase of a pest population after a period of decline, to a level higher than its original one, especially when the decline is caused by a pest control measure

Retention period : the time following acquisition feeding during which a vector is capable of transmitting virus.

Rhizosphere : the soil closely surrounding a living root and influenced by this as substrate of natural microflora; the zone of root influence; the term refers to the thin layer adhering to a root after the loose soil and clumps have been removed by shaking

Ribosomal RNA : molecules forming part of the ribosomal structure

Saprophyte : organisms (usually fungus) living on dead and decaying tissues.

Satellite : a term first used to describe a small virus (satellite virus) associated with tobacco necrosis virus (TNV), which is dependent upon the TNV genome for its own replication. Also used to describe certain nucleic acid molecules that are unable to multiply in the host cell without the aid of other nucleic acid molecules.

Satellite DNA : DNA in eukaryotes that is present in large numbers of repetitive copies

Sclerotium : a long lived, hard compact mass of vegetatively produced fungal hyphae, usually black on the outside, serving as a resting stage from which fructifications may be formed

Secondary infection : any infection caused by inoculum produced as a result of primary or a subsequent infection ; an infection caused by secondary inoculum

Sedimentation coefficient : the rate of sedimentation of virus per unit centrifugal field measured in Svedberg units

Semi persistent transmission : virus transmission by an insect vector that is intermediate between non-persistent and persistent transmission

Serology : a method using the specificity of the antigen-antibody reaction for the detection and identification of antigenic substances and the organism that carry them (e.g. micro-organisms or viruses).

Serotype : a subgroup of a particular species unique because of the anti-genic properties of some cellular component; a given species may contain many different subgroups of different serotype

Slime layer : a diffuse layer of polysaccharide exterior to the cell wall in some bacteria

Soil inhabitant : an organism that maintains its population in soil over a period of time.

Soil solarization : a hydrothermal process that occurs in moist soil which is covered by a transparent or black plastic film and exposed to sunlight during warm summer months. During solarization soil temperatures are increased to lethal level for many plant pathogens and pests but which can improve the growth and development of plants

Somaclonal variation : somatic (vegetative non-sexual) plant cells can be propagated *in vitro* in an appropriate nutrient medium. According to the composition and conditions the cells may proliferate in an un-differentiated (disorganized) pattern to form a callus or in a differentiated (orgnised) manner to form a plant with a shoot and root. The cells which multiply by division of the parent somatic cells are called somaclones and, theoretically, should be genetically identical with the parent. In fact *in vitro* cell culture of somatic cells, whether from a leaf, a stem, a root, a shoot or a cotyledon, frequently generates cells significantly different, genetically, from the parent. During culture the DNA breaks up and is reassembled in different sequences which give rise to plants different in identifiable characters from the parent. Such progeny are called somaclonal variants and provide a useful source of genetic variation

Somaclone : plants produced by a genetic engineering technique by which single cells or protoplasts are cultured to produce individuals which are genetically variable from their genetically stable parent. The variation induced is called somaclonal

Somatic embryogenesis : the generation from somatic cell or tissue culture of bipolar embryos, similar to sexually derived embryos. Both sexual and somatic embryos possess a primordial root and shoot

Somatic hybridization : the formation of hybrids by fusion of somatic cells, as opposed to the fusion of gametes. The term is commonly applied to fusion of plant protoplasts

Southern blotting : this is a method for transferring separated DNA fragments from an agarose gel to a solid support such as a nitrocellulose membrane

Spectrophotometer : an instrument that analyses characteristics of organic compounds through transmittance of light waves through solutions. Ultra-violet, and infra-red wave lengths are commonly employed. Ordinary light can be used in certain types of instruments, and electromagnetic fields are employed in some instruments

Spermagonium : structure in which male reproductive cells are produced; in rust fungi, globose or flask-shaped haploid fruiting body composed of receptive hyphae and spermatia (pycniospores)

Sporodochium : a closely woven mycelial stroma in which conidiophores develop over the surface of an erumpent, cushion-like fungal structure

Steam sterilization : treatment of soil with steam to depress growth of harmful organisms

Sterilization : the total destruction of living organisms by various means, including heat, chemicals or irradiation

Streak plate method : a procedure for separating cells on a sterile agar surface so that individual cells grow into distinct and separate colonies.

Stylet-borne transmission : a type of virus transmission in which the virus is acquired and transmitted by the vector after short feeding times, and is retained by the vector for only a short period of time.

Symptom : a visible abnormal change in a host as a result of pest infestation or infection

Symptomless carrier : a plant that, although infected with a pathogen (usually a virus), produces no obvious symptoms

Syndrome : the totality of effects produced in a host by one disease, whether simultaneously or successively and whether detectable to the unaided eye or not

Totipotency : the concept that even specialized cells contain all of the genetic information for an organism and therefore, any cell should be able to regenerate into any tissue or into an entire plant

Toxicity : capacity of a substance to interfere with the vital processes of an organism.

Transcription : the process of RNA synthesis by RNA polymerase to produce a single-stranded RNA complementary to a DNA template

Transduction : the transfer of genes from one organism to another by viruses, especially in bacteria

Transformation : a genetic engineering procedure whereby a piece of foreign DNA is transferred to a cell thus conferring upon its novel characters. Also, the change of a normal to a malignant cell

Transgenic : an organism whose genome has been altered to include new genes from the same or different species

Translocation : movement of water, nutrients, chemicals, or food materials within a plant.

Transovarial transmission : a mode of transmission in which the passage of micro-organisms from mother to egg is known to occur within the ovary. Transmission of a virus from an adult organism to its progeny through the ovaries and eggs.

Ultra-centrifuge : a centrifuge with an extremely high rate of rotation which will separate out very small particles

Ultraviolet spectrophotometry : a technique in the identification of organic compounds by passing different wave lengths of the ultra-violet region of light through solutions in suitable solvents

Uredospore : a binucleate, repeating spore (II state) of the Uredinales.

Variegation : pattern of two or more colors in a plant part, as in a green and whit leaf

Vector : a living organism able to carry and transmit a pathogen and disseminate disease; (in genetic engineering) a vector or cloning vehicle is a self-replicating DNA molecule, such as a plasmid or virus, used to introduce a fragment of foreign DNA into a host cell.

Vegetative compatibility group (VCG) : refers to strains that can form a stable vegetative heterokaryon, implying identity of alleles at every vic locus. This usually employed with *Fusarium* sp (Leslie, 1993).

Vegetative propagation : asexual reproduction ; implants, the use of cuttings, bulbs, tubers, and other vegetative plant parts to grow new plants.

Vertifolia effect : proposed by Van der Plank it is the decline of horizontal resistance due to the absence of selection pressure, during breeding for vertical resistance.

Vesicular-arbuscular mycorrhiza : symbiotic association between a nonpathogenic or weakly pathogenic fungus and the roots of plants in which fungal hyphae invade cortical cells of the root and produce vesicles and arbuscles.

Victorin : victoxinine, a strong pathotoxin produced by *Dreschlera victories* in *Avena sativa.*

Vigour : physiological potential for rapid and uniform germination and fast seedling growth under ordinary field conditions.

Viricide : a substance that completely and permanently inactivates a virus

Virulence : the degree of ability to produce disease; often used to denote qualitative rather than quantitative differences of pathogenicity.

Virulent : capable of producing disease.

Viruliferous : containing a virus; of an insect vector, containing virus and being capable of introducing it into a suscept

Virus, circulative : originally restricted to viruses that have been proved to be transmitted by way of the salivary glands of leafhoppers. Such viruses can be detected in the haemolymph and persist for a long time in the vector and through the moult. The term was extended by J.S.Kennedy, M.F. Day & V.F. Eastop to include additional viruses which are assumed to be transmitted in this way be leafhoppers and other types of vector. In this sense the circulative viruses are equivalent to the internally transported viruses of M.A.Watson. Some circulative viruses are known to multiply in the vector and have been referred to as circulative propagative or simply as propagative viruses. Note, however, that circulative and propagative refer to different attributes of the virus and are not synonymous; cf. virus, stylet-borne.

Virus, stylet-borne : one which has been proved to be transmitted on the stylet of its vector. Such viruses fail to pass through the moult or to appear in the haemolymph and do not persist long in the vector. The term was extended by Kennedy et al. (1962) to include many other viruses which have been assumed to be carried on the stylets.

Virus : a submicroscopic, intracellular, obligate parasite consisting of a core of infectious nucleic acid usually surrounded by a protein coat

Virus, helper : one which must be present for the transmission of a second virus by its vector. Synonymous with assistor virus, carrier virus

Virus, non-persistent : one which is retained for not more than a few hours by vectors transferred from infected to healthy plants. Such viruses share additional transmission characteristics; they fail to pass to the progeny of the vector or through the moult; they seem to be carried externally on the stylets and do not multiply within the vector or appear in the haemolymph; they are acquired most readily by starved vectors after very short acquisition feeding periods; they have very short transmission threshold periods. These additional characteristics should not be regarded as criteria for applying the term non persistent.

Virus, satellite : one which multiplies only in the presence of a specific second virus.

Virus, semi-persistent : one which is retained for a maximum of several days by vectors transferred from infected to healthy plants. The transmission characteristics of this small group of viruses are less clearly defined than those of the non-persistent and persistent viruses. The efficiency of transmission increases with increases in both acquisition and inoculation feeding times. There is no latent period and the greatest likelihood of transmission is immediately after vectors leave the virus source. Pre-acquisition fasting has no effect on the efficiency of transmission. The semi-persistent viruses do not pass to the progeny of the vector or appear in the haemolymph and thy do not seem to pass through the moult.

Virus, NETU :   a term introduced by Cadman (1963) for nematode-transmitted viruses with tubular particles. These viruses comprise the tobra virus group of Harrison et al.

Virus inhibitor: a chemical that prevents virus transmission without inactivating the virus. Inhibition of transmission may be temporary and can sometimes be avoided by dilution. Inhibitors frequently act on the recipient host rather than on the virus itself.

Virus transmission : the transfer of a virus from an infected host to a non-infected one, commonly through the agency of another organism, the vector, which is often an insect or a nematode, but also may be a mite, fungus, parasitic flowering plant etc. Transmission may also be effected by direct contact, e.g. by rubbing together of leaves or stems, through root grafts etc. The terminology of transmission related both to the properties of the virus and to its relations with the vector. Hence, the virus may be persistent, semi-persistent or non-persistent, according to the length of time that it remains infective in the vector; propagative, if it multiplies within the vector, or non propagative; stylet-borne dependent on the presence of a second virus, or not dependent

Virus, dependent : one which is transmitted by its vector only in the presence of a second (helper) virus. Synonymous with assisted virus, carried virus, helped virus

Virus, ILAR : a term introduced by R.W. Fulton, for Prunus necrotic ringspot, and similar viruses with isometric particles that are labile and cause ring-spot symptoms in at least some of their hosts.

Virus, multiparticulate : one which produces more than one type of nucleoprotein component in infected plants

Vivotoxin : a toxin produced by a parasite or by a host, in response to a parasite, or both, that operates in causation of a disease not being itself the primary agent.

Yellows : a disease type characterized by pronounced chlorosis.

On-Line Resources

IPM	www.pestmanagement.co.uk/library/glossary.html
BSPP	www.bspp.org.uk
APSnet	www.apsnet.org/education/illustrated
Biocontrol	www.googlesearch.com
Biotechnology	www.plpa.agri.umn.edu

References

Adams, A.N. and Barabara, D.J. (1982), the use of F (ab) 2-based ELISA to detect serological relationships among carlaviruses. *Ann. Appl. Biol.* 101: 495-500.

Akram, Mohd., Jain, R.K.Choudary, Vikas, Ahlawat, Y.S. and Khurana, S.M.Paul (2004). Comparison of groundnut bud necrosis virus isolates based on movement protein (NSM) gene sequences. *Ann.Appl. Biol.* 145 : 285-289.

Alvarez, A.M., Benedict, A.A., Or, G. and Mizumoto, C.Y.1987. Identification of Xanthomonads from crucifer seeds with monoclonal antibodies. *Phytopathology* 77: 1725 (Abstr).

Anne M. Alvarez, 2004. Integrated approaches for detection of plant pathogenic bacteria and diagnosis of bacterial diseases, Annu, Rev. *Phytopathol*. 42: 339-66.

Anonymous (1980) Survey for the presence of mycotoxins with special reference to patulin, steringmatoxystin and penicillic acid in foods, food products, animal feeds and concentrates available in the Tamil Nadu region of India. Final Technical Report (DST). Dept. of Food Technology, Tamil Ndu Agric. University, Coimbatore. (1).

Anonymous (1989) Survey for the presence of mycotoxins with special reference to patulin, steringmatoxystin and penicillic acid in foods, food products, animal feeds and concentrates available in the Tamil Nadu region of India. Final Technical Report (DST), Dept. of Food Technology, Tamil Nadu Agric. University, Coimbatore. (1).

Anonymous (1998). *Phytopathora* Genome initiative (PGI) White Paper: http://www.nogr.org/research/pgi/whitepaper. html.

Ash, C.Farraw, A.E, Dorsch, M., Stackbrandt, E. and Collins, M.D. (1991). Comparative analysis of *Bacillus anthracis*, *Bacillus cereus*, and related species on the basis of reverse transcriptase sequencing of 16S rRNA. *Int. J.Syst. Bacteriol*. 41: 343-46.

Ball, E.M. and Brakke, M.K. (1968). Leaf-dip serology for electron microscopic identification of plant viruses. *Virology* 36: 152-155.

Banttari, E.E. and Khurana, S.M. Paul (1998). Serological procedure in Plant Virology. In: Pathol Problems of Econ. Crop Plants & Their Magmt. (Ed. Khurana, S.M.Paul) pp 603-624, Scientific Publishers, Jodhpur.

Banttari, E.E. and P.H. Goodwin. 1985. Detection of potato viruses S, X and Y by enzyme-linked immunosorbent assay on nitrocellulose membranes. *Plant Dis.* 69 : 202-205.

Beachy RN, Loesch-Fries S and Tumer N E, 1990. Coat-protein-mediated resistance against virus infection. *Annu Rev Phyhtopathol*, 28, 451-474.

Bendahmane A, Kanyuka K and Baulcombe D C, 1999. The Rx gene from potato controls separate virus resistance and cell death responses. *Plant Cell*, 11, 781-791.

Bendahmane A, Querci M, Kanyka K & Baulcombe D C, 2000. Agrobacterium transient expression system as a tool for the isolation of disease resistance genes: application to the Rx2 locus in potato. Plant, 21, 73-81.

Bhat A I, Jain R K, Varma A, Chandra N and Lal S K, 2001b. Tospovirus (es) infecting grain legumes in Delhi-their indentification by serology and nucleic acid hybridization. *Indian Phytopath*, 54: 112-116.

Bhat AI, Jain R K, Kumar A, Ramaiah M and Varma A, 2001 a. Serological and coat protein sequence studies suggest that necrosis disease on sunflower in India is caused by a strain of Tobacco streak virus. Arch Virol, (in Press).

Bhat, R.V. (1991). Aflatoxins: Successes and failures of three decades of research. Pages 80-85. In Fungi and Mycotoxins in stored products. ACIAr proceedings No.36 (Eds). B.R. Champ, E.Highley, A.d. Hocking an J.I. Pitt. The Griffin Press Ltd., Netley, South Australia (2).

Bilgrami, K.s. Sahay, S.S., Shrivastava, A.K. and Rahman, M.F. (1990) incidence of zearalenone, DON and T-2 toxin producing strains of Fusarium sp on food items. *Proc. Indian Nat. Sci. Acad.* B56 : 223-228 (5).

Bilgrami, K.S. Sinha K.K. and Singh, A. (1983). Chemical changes in dry fruits during aflatoxin elaboration by *Aspergillus flavus Link ex Fries*. *Curr. Sci.52* : 960-964 (3).

Biondi S, Miszu J, Miltempergher L and Baghi N 1991 Selection of elm cell culture variants resistant to *Ophiostoma ulmi* culture filtrate. *Journal of Plant Physiology* 137 (5): 631-634.

Birnboim, H.C. and Doly, J. 1979. A rapid alkaline extraction procedure for screening recombinant plasmid DNA. *Nucleic Acids Research* 7 : 1513-23.

Blaney, B.J. (1991). *Fusarium* and *Alternaria toxins*. Pages 88-98. In : Fungi and mycotoxins in stored products. ACIAR proceedings No.36 (Eds. B.R.Champ, E.Highley, A.D. Hocking and J.I. Pitt. The Griffin Press Ltd., Natley South Australia (6).

Boonham, N., Smith, P., Walsh, K., Tame, J. and Morris, J. (2002a). The detection of Tomato spotted wilt virus (TSWV) in invidual thrips vectors using real-time fluorescent RT-PCR (TaqMan). *J.Virol. Methods* 101: 37-48.

Boonham, N., Walsh, W., Preston, S., North, J. and Smith, P. (2002b). The detection of tuber necrotic isolates of Potato virus Y, and the accurate discrimination of PVY(O), PVY (N) and PVY (C) strains using RT-PCR. *J.Virol. Methods* 102:103-112.

Braynt JA 1974 Molecular aspects of differentiation. In cell Cycle Controls (ed JA Braynt) Academic Press. New York.

Briard, M., Dutertre, M., Rouxel, F. and Brygoo, Y. (1995). Ribosomal RNA sequence divergence within the Pythiaceae. *Mycol. Res.*99:1119-1127.

Bullerman, L.B. (1986). Mycotoxins and food safety. A specific status summary by the Insittute of Food Technologist's expert panel on food safety and nutrition. Institute of Food Technologiests. *Chicago* (8).

Bullerman;, L.B. (1979). Significance of mycotoxins to food safety and human health. J. *Food. Prot.* **42**: 65-85 (7).

Burdaspal, P., and Pinilla. I. (1979) *Alimentaria*. 107: 35-37 (9).

Cahill, D.M. and Hardham, A.R. (1994). Exploitationof zoospore taxis in the development of a novel dipstick immunoassay for the specific detection of Phytophthora cinnamomi. *Phytopathology* 84:193-200.

Carbone, I., Anderson, J.B. and Kohn, L.M. (1999). Patterns of descent in clonal lineages and their multilocus fingerprints are resolved with combined gene genealogies. *Evolution* 53:11-21.

Carlton, W.W. Sansing, G. Szczech. G.M. and Tuite, J. (1974). Citrinin mycotoxicosis in beagle dogs. Food Cosmet. *Toxicol.* 12: 479-440. (10).

Carlton, W.W. Sansing, G.Szczech. G.M. and Tuite, J. (1974). Citrinin mycotoxicosis in beagle dogs. *Food Cosmet. Toxicol.* **12** : 479-440. (10).

Chu, P.W.G., Waterhouse, P.M., Martin, R.R. and Gerlach, W. W. (1989). New approaches to the detection of microbial plant pathogens. Biotechnol. *Genet. Eng. Rev.* 7:45-111.

Clark, M.F. (1981). Immunosorbent assays in plant pathology. *Annu.Rev.Phytopath.* 19:83-106.

Clark, M.F. and Adams, A.N. (1977). Characteristics of the microplate method of enzyme-linked immunosorbent assay for the detection of plant viruses. *J.Gen. Virol.* 34: 475-483.

Claugher, D. 1990. Scanning Electron Microscopy in taxonomy and functional morphology. Clarendon Press, Oxford.

Cockerill, F.R. and Smith, T.F. (2002). Rapid-cycle real-time PCR: revolution for clinical microbiology. *Am.Soc. Microbiol. News* 68 : 77-83.

Coddington A, Matthews, P.M., Cullis, C. and Smith, K.H. (1987). Restriction digest patterns of total DNA from different races of *Fusarium oxysporum* fo. sp. *pisi* an improved method for race classification. *J. Phytopathol.* 118:9-20.

Cooke, DEL., Kennedy, D.M., Guy, D.C., Russell, J., Unkles, S.E. and Duncan, J.M. (1996). Relatedness of Group I species of *Phytophthora* as assessed by randomly amplified polymorphic DNA (RAPDs) and sequences of ribosomal DNA. *Mycol. Res.* 100: 297-303.

Cooper B, Lapido M, Heick JA, Allan Dodds J and Beachy R N, 1995. A defective movement protein of TMV in transgenic plants confers resistance to multiple viruses whereas the functional analog increases susceptibility. *Virology*, 206, 307-313.

Crawford, A.R., Bassam, B.J., Drenth, A., McLean, D. J. and lrwin, J.A.G. (1996). Evolutionary relationships among *Phytophthora* species deduced from rDNA sequence analysis. *Mycol. Res.* 100:437-443.

Dasgupta, I., Malathi, V.G. and Mukherjee, S.K. 2003. Genetic engineering for virus resistance. *Curr Sci.*, 84: 341-354.

DeBoer, S.H. and McNaughton, M.E. (1986). Evaluation of immunofluoscence with monoclonal antibodies for detecting latent bacterial ring rot infections. *Am. Potato* J. 63: 553-543.

DeBoer, S.H. and MCNaughton, M.E. (1987). Monoclonlal antibodies to the lipopolysaccharides of Erwinia caratovora sub sp.atroseptica serogroup 1. Phytopathology 77: 828-32.

Dehne, H.W., Adam, G., Diekmann, M., Frashm, J., Machnik, A.M. and Halteren, P.V. (1997). Diagnosis and identification of plant pathogens. In:4th int. Plant Pathol Symp., Eur. Found., Kluwer, Dordrecht, Netherands.

Derrick, K.S.(1973). Quantitative assay for plant viruses using serologically specific electron microscopy. *Virology* 56: 652-653.

Dhar, A.K. and Singh, R.P. (1994). Improvement in the sensitivity of PVY[N] detection by increasing the cDNA probe size. *J.Virol. Methods* 50: 197-210.

Didenko, V.D. (2001). DNA probes using fluorescence resonance energy transfer (FRET): designs and applications. *Biotechniques* 31:1106-1120.

Ding, S.W., Howe, J., Keese, P., Mackenzie, A. and Meek, D. (1990). The tymobox, a sequence shared by most tymoviruses: its use in molecular studies of tymoviruses. *Nucleic Acids Res.* 18: 1181-87.

Doolittle, R.F. (1998). Microbial genomes opened up Science 392:339-342.

Doyle. J.J. and Doyle J.N. 1987. *Phytochem Bull.* 19 : 11-15.

Duncan, J.M. and Torrance, L. (1992). Techniques for the Rapid Detection of Plant Pathogens. Blackwell sci., Oxford.

Ersek, T., Schoeiz, J.E. and English, J.T. (1994). PCR amplification of species-specific DNA sequences can distinguish among *Phytophthora* species. *Appl. Environ. Microbiol.* 60: 2616-2621.

Eun, A.J.C. and Wong, S.M. (2000). Molecular beacons: a new approach to plant virus detection. *Phytopathology.* 90:269-275.

Fitchen JH and Beachy RN, 1993. Genetically engineered protection against viruses in transgenic plants. Annu Rev Microbial, 47, 739-763.

Forster, R.L. (1999). Ground surveillance. http://www. Apsnet. org/ online/ feature/biosecurity/abstracts # htm # forster.

Francone R, Roggero P, Perazzi P, Arias F J, Desideru A, Bittz O, Pashkouler D, Master B, Bracee L, Masenga V, Milne R G and Benveruto E, 1999. Functional expression in bacteria and plants of an scFv antibody fragment against tospoviruses. *Immunotechnology* 4, 189-201.

Frederick, R.D., Snyder, C.L., Peterson, G.L. and Bondle, M.R. (2002). Polymerase chain reaction assays for the detection and discrimination of the soyabean rust pathogens *Phakopsora pachyrhizi* and *P.meibomiae. Phytopathology*. 92 : 217-227.

Frederick, R.D., Snyder, K.E., Tooley, P., Berthier – Schaad, Y. and Peterson, G.L. (2000). Identification and differentiation of *Tilletia indica* and *Tilletia walkeri* using PCR. *Phytopathology* 90: 951-960.

Frisvad, J.C. (1987). High performance liquid chromatographic determination of profiles of mycotoxins and other secondary metabolities. *J.Chromatogr.* 392 : 222-347. (12).

Frolova LV and Shamina ZB 1974 The cytogenetic characteristics of tissue culture in plants of leguminosae. Tsitologica Genetica 8 : 413-418.

Fry. W.E. (2001). Molecular tools in epidemiology of plant diseases with special reference to potao *Phytophthora infestans* pathosystem. Plant Dis. Res.16: 1-9.

Ftichen JH and Beachy RN, 1933. Genetically engineered protection against viruses in transgenic plants. *Annu. Rev Microbial*, 47, 739-763.

Fuchs M, Ferreira S & Gonsalves D, 1997. Mangement of virus diseases by classical and engineered protection. Molecular Plant Pathology on line 1997 /0116 fuchas..

Gamborag OL, Constable F and Shyluk J P 1975, Organogenesis in callus from shoot apices of *Pisum sativum. Physiologia Plantarum* 30: 125-128.

Gamborg O L Shyluk J and Kartha K K 1975 Factors affecting the isolation and callus formation in protoplastics from the shoot apices of *Pisum sativum.* Plant Science Letters 4 : 285-299.

Gamborg O L, Miller R A and Ojima K 1968 Nutrient requirements of suspension cultures of soybean root cells. *Experimental Cell Research* 50: 151-158.

Garg, I.D. and Khurana, S.M. Paul (1992). Factors influencing immune electron microscopy of flexuous potato viruses. *Acta Virologica* 36 : 435-442.

Garg, I.D. and Khurana, S.M. Paul (1993). ISEM dislodging of PVY/PVA virions upon incubation with antisera to PVS/PVM. *J.Indian Potao Assoc.* : 20-80.

Garg, I.D., Hedge, V. and Khurana, S.M. Paul (2000). Effect of pH of antisera, bovine serum albumin and ions on the stability of immunosorbent flexuous potato viruses. *Acta Virologica* 53: 256-260.

Gautheret R J 1934 Culture du tissue cambial C R Hebd. *Seances Academic Science* 198: 2195-2196.

Gautheret R J 1966 Factors affecting differentiation of plant tissues grown in vitro. pp. 55-95. In cell differentiation and morphogenesis. Amsterdam. North Holland.

Geetha N, Venkatachalam P and Rao G R 1997 *In vitro* selection and characterization of PEG-tolerant callus lines of *Vigna mungo* L. Hepper. *Current Agriculture* 20(1-2): 77-82.

Gerlach WL, Llewellyn D & Haseloff J, 1987. Construction of a plant disease resistance gene from the satellite RNA of tobacco ringspot virus. *Nature Lond*, 328, 802-805.

Gibbs, A. and Mackenzie. A. (1997). A primer pair for amplifying part of the genome of all potyvirids by RT-PCR. *J.Virol. Methods* 63L 9-16.

Gonsalves D, 1998. Control of papaya ringspot virus in papaya: a case Study. *Annu Rev Phytopathol,* 36, 415-437.

Goodwin, P.H., English, J.T., Neher, D.A., Duniway, J.M. and Kirkpatrick, B.C. (1990a). Detection of *Phytophthora parasitica* from soil and host tissue with a species-specific DNA probe. Phytopathology 80: 277-281.

Goodwin, P.H., Kirkpatrick, B.C. and Duniway, J.M. (1990b). Identification of *Phytophthora citrophthora* with cloned DNA probes. Appl. Environ. Microbial. 56: 669-674.

Goodwin, P.H., Kirkpatrick, B.C. and Duniway, J.M.(1989). Cloned DNA probes for identification of *Phytophthora parasitica*. Phytopathology 79: 716-721.

Goodwin, S.B. and Drenth, A. (19970. Origin of the A2 mating type of *Phytophthora infestans* outside Mexico. Phytopathology 87: 992-99.

Goodwin, S.B., Spielman, L.J., Matuszak, J.M., Bergerson, S.N. and Fry, W.E. (1992). Clonal diversity and genetic differentiation of *Phytophthora infestans* populations in northern and central Mexico. *Phytopathology* 82: 955-61.

Gray L E, Guan Y G and Widholm J M 1986 Reaction of soybean callus to culture filtrates of *Phialophora gregata*. *Plant Science* 47 (1): 45-55.

Grove, J.F. (1988). Non-macrocyclic trichotecenes. Natural Products Reports 5: 187-209. (13)

Grumet R, 1994. Development of virus resistant plants via genetic engineering. *Plant Breed Rev*, 12, 47-49.

Gundersen, D.E., Lee, I.M., Rehner, S.A., Davis, R.E. and Kingbury, D.T. (1994). Phylogeny of mycoplasma-like organisms (Phytoplasmas): a basis for their classification. *J.Bacterial.* 176: 5244-5254.

Haberlandt G 1902 Kulturuesuche mitisolierten P flanzen.   Sizungzer. Matt-Naturwiss K L Kaiser A Kad Wiss Wien 111 : 69-92.

Hames, B.D. 1998. Gel electrophoresis of Proteins : A Practical Approach. Oxford University Press, pp 352.

Hamilton R I, 1980. Defenses triggered by previous invaders : viruses. In Plant Disease, 5th ed. Academic Press, New York, Pp. 279-303.

Hampton, R.,E. Ball and De boer S. 1990. Serological methods for detection and identification of viral and bacterial plant pathogens – a laboratory manual. APS press, Minnesota, USA, 389 pp.

Hardham, A.R., Suzaki, E. and Perkin, J.L. (1986). Monoclonal antibodies to isolate, species, and genus-specific components on the surface of zoospores and ysts of the fungus *Phytophthora cinnamomi*. *Can.J.Bot.* 63: 311-321.

Harrison B D, Mayo M A & Baulcombe D C, 1987. Virus resistance in transgenic plants that express cucumber mosaic virus satellite RNA. *Nature Lond*, **328**, 799-801.

Harwig, J. Scott. P.M. Slotz. D.R. and Blanchfield, B.I. (1979). Toxins of molds from decaying tomato fruits *Appl. Environ. Microbiol.* **38**: 267-274 (14).

Hawkes, R.,Niday, E. and Gordon, J.(1982). A dot immuno-binding assay for monoclonal and other antibodies. *Analytical Biochem.* 119:142-147.

Henson, J.M. and French, R. (1993). The polymerase chain reaction and plant disease diagnosis. *Annu.Rev. Phytopath*. 31. 81-109.

Hiatt, A., Cafferkey, R & Bowdish, K. 1989. Production of antibodies in transgenic plants. *Nature* 342: 76-78.

Holland, P.M., Abramson, R.D, Watson, R. and Gelfand, D.H. (1991). Detection of specific polymerase chain reaction product by utilizing the 5' to 3' exonuclease activity of Thermus aquaticus DNA polymerase. *Proc.Nati. Acad. Sci.* USA 88: 7276-80.

Hsiech, D.P.H. Lin, M.T. and Yao. R.C. (1973). Conservation of sterigmatocystin to aflatoxin B1. *Bichem. Biophys. Res. Comm.* **52**: 992-997 (15).

Hsu, Y.H. (1984). Immunogold for detection of antigen on nitrocellulose paper. *Annal. Biochem.* 142: 221-25.

Hull, R. 2002. Matthew's Plant Virology. Academic Press, New York.

Jacobi V., Bachand G.D., Hamelin R.C. and Castallo J.D. (1998). Development of a multiplex immunocapture RT-PCR assay for detection of tomato and tobacco mosaic tobamoviruses. *J.Virol. Methods* 74: 167-178.

Jain, R.K. Khurana, S.M. Paul, Bhat, A.I. and Choudhary, Vikas (2004a). Nucleocapsid protein gene sequence studies confirm that potato stem necrosis disease is caused by a strain of groundnut bud necrosis virus. *Indian Phytopath.* 57: 167-173.

Jain, R.K., Sharma, J., Sivakumar, A.S.,Sharma, P.K., Byadgi, A.S., Verma, A.K. and Varma, A. (2004b). Variability in the coat protein gene of papaya ringspot virus isolates from multiple location in India. *Arch. Virol.* 149: 2435-2442.

James, D., Jelkann, W. and Upton, C. (1999). Specific detection of cherry mottle leaf virus using digoxigenin-labeled cDNA probes and RT-PCR. *Plant Dis.* 83: 235-239.

Jarvis, B.B.(1898). Mycotoxins an overview, pages 18-29. In: Natural toxins characterization, pharmacology and tehrapeutics. Proceedings of the 9th World Congress on Animal, Plant and (Eds, Microbial Toxins C.L. Ownby, and G.V.Odell), Pergamon Press, New York. (18).

Jelinek, C.F. (1987). Distribution of mycotoxins and analysis of worldwide commodities data, including data from FAO/WHO/UNEP food contamination monitoring programme. Joint FAO/WHO/UNEP Second International Conference on Mycotoxins Bangkok, Thailand, September 28 to October 3, 1987. (19).

Jemmali, M and Mazenand, C. (1980) *Ann. Microbiol* (Paris) 131B: 319-321 (20).

Jin H, Hartman G I, Huang Y H, Nickell C D and Widholm J M 1996 Regeneration of soybean plants from embryogenic suspension cultures treated with toxic culture

Joan M. Herson; and Roy French. 1993. The polymerase chain reaction and plant disease diagnosis. *Annu Rev. Phytopathol.* 31 : 81 – 109.

Johannsson, A, Stanley, C.J.and Self, C.H. (1985). A fast highly sensitive colorimetric enzyme immunoassay system demonstrating benefits of enzyme amplification in clinical chemistry. *Clinical. Chemica. Acta.* 148: 119-124.

Karasev, A.V., Nikolaeva, O.V., Koonin E.V., Gumpt, D.J. and Garnsey S.M. (1994). Screening of the closterovirus genome by degenerate primer medicated polymerase chain reaction. *J.Gen. Virol.* 75. 1415-1422.

Khurana, S.M. Paul (2004). Potato viruses and their management. pp. 389-440. In: Diseases of Fruits and Vegetables: Diagnosis and Management

Vol.II. (SAMH Naqvi Ed.). Kluwer Academic, Dordrecht, Boston and London.

Khurana, S.M. Paul and Garb, I.D. (1993). New techniques for detection of viruses and Viroids pp. 529-566. In: Advances in Horticulture Vol.7 (K.L.Chandha and J.S.Grewal eds.). Malhotra Publishing House, New Delhi.

Kimball S L and Bingham E T 1973 Adventitious bud development of soybean hypocotyls sections in culture. *Crop Science* 13: 758-760.

Kistler, H.C., Bosland, P.W., Benny, U., Leong, S. and Williams, P.H. (1987). Relatendness of strains of Fusarium oxysporum from crucifers measured by examination of mitochondrial and ribosomal DNA. *Phytopathology* 77: 1289-93.

Klerks, M.M., Leone, G.O., Verbeek, M., Van den Heuvel, J.F. and Schoen, C.D. (2001). Development of a multiplex amplified DetRNA for the simultaneous detection of Potato leafroll virus and Potato virus Y in potato tubers. *J. Virol. Methods* 93: 115-25.

Koening, R. and Paul, H.L. (1982). Variants of ELISA in plant virus diagnosis. *J. Virol. Methods* 5: 113-125.

Korimbocus, J., Coates, D., Barker, I. And Boonham, N. (2002). Improved detection of Surgacane yellow leaf virus using a real-time fluorescent (TagMan) RT-PCR assay. *J.Virol.* Methods 103: 109-120.

Kotee W 1922 Kultuersche mitisolierten W vrzelspitzen Bietr. *Allgen Botan* 2 : 413-434.

Kumar MR and Balasubramanian K.A. 2007. Genetic relatedness, susceptibility and resistant levels of blackgram genotypes to leaf spot fungus, *corynespora Cassicole. I Mycol Pathol* 37 (1): 101 – 104.

Koon-Hui Wang and Robert Mc Sorley (2003). Beneficial soil borne fungi on line publication.

Kumar A 1974 Vitamin requirements of callus tissues of *Arachis hypogea* L. Indian Journal of Experimental Biology 12: 465-466.

Lafferty, K.J. and Oertelis, S.J.(1961). Attachment of antibody to influenza virus. *Nature* (London) 192: 764-765.

Lamontagne B, Larose S, Boulanger J & Elela S A, 2001. The Rnase III family : a conserved structure and expanding functions in eukaryotic ds RNA metabolism. Curr isues Mol Biol, 3, 71-78.

Leach, J.E. (1999). Assuring food security: detecting and controlling modified pathogens. http://www.aspnet. Org/online/feature/biosecurity /top.htm.#leach.

Le-Banna, A.A. Scott, P.M. Law, P.Y.Sakuma, T.Platt, H.W. and Campbel, V. (1984). Formation of trichothecenes by *Fusarium solani* var. *coeruleum* and *fusarium sambucimum* in potatoes. *Appl. Environ. Microbiol.* **47**: 1160-1171. (11).

Lee, I.M., Davis, R.E. and Gundersen, D.E. (2000). Phytoplasma: Phytopathogenic molicutes. *Annu. Rev. Microbiol.* 54: 221-255.

Lee, I.M., Gundersen, D.E., Hammond, R.W, and Davis, R.E. (1994). Use of mycoplasmalike organism (MLO) group-specific oligonucleotide primers for nested-PCR assays to detect mixed MLO infections in a single host plant. *Phytopathology* 84: 559-66.

Lee, S.B., White, T.J.and Taylor, J.W. (1993). Detection of Phytophthora species by oligonucleotide hybridization to amplified ribosomal DNA spacers. *Phytopathology* 83: 177-81.

Lim, P.O. and Sears, B.B. (1989). 16S rRNA sequence indicates that plant-pathogenic mycoplasma-like organisms are evolutionarily distinct from animal mycoplasmas. *J.Bacteriol.* 171: 5901-5906.

Lindquist, H., Koponen, H. and Valkonen, J.P.T. (1998). Peronospora sparsa on cultivated Rubus arcticus and its detection by PCR based on ITS sequences. *Plant Dis.* 82: 1304-1311.

Lodge J, Kaniewski W K and Tumer N E, 1993. Broad-spectrum virus resistance in transgenic plants expressing pokeweed antiviral protein. *Proc Nat Acad Sci* USA, 90, 7089-7093.

Lomonossoff G P, 1995. Pathogen-derived resistance to plant viruses. *Annu Rev Phytopathol*, 33, 323-343.

Lounis; F.J., Rademaker, J.L.W., and de Bruij, F.J.1999. The three Ds of PCR - based genomic analysis of phytobacteria: Diverisity, detection and diagnosis. *Annu. Rev. Phytopathol.* 37: 81-125.

Majer, D., Mithern, R., Lewis, B.G., Vos, P. and Oliver, R.P. 1996. The use of AFLP finger printing for detection of genetic variation in fungi. *Mycological Research.* **10** : 1107-1111.

Malathi, V.G., Rashmi Aggarwal., Jayashree Jayaraman., Bhat, I.G; and Singh, D.V. 2001. Applications of molecular techniques in Plant Pathology - A practical manual. IARI, New Delhi.

Manicom, B.Q., Bar-Joseph, M., Rosner, A., Vigodsky Haas, H. and Kotze, J.M. (1987). Potential application of random DNA probes and restriction fragment length polymorphisms in the taxonomy of the Fusaria. *Phytopathology* 77: 669-672.

Martin R.R., James, D. and Levesque, C.A. (2000). Impacts of molecular diagnostic technologies on plant disease management. *Annu. Rev. Phytopath.* 38 : 207-239.

Maslennikov S E, Posypanov G S and Mazentsev A V 1994 Developing methodological approaches for obtaining forms of Lucerne resistant to *Fusarium* using callus and cell cultures. Izvestiya-Timiryazevski-Sel Skokhozyaistvehnoi-Akademi 3: 177-186.

Mathews REF, 1999. Plant Virology Academic Press, New York.

Maxam, A.M. and Gilbert, W. 1977. A new method for sequencing DNA. *Proc. Natl. Acad. Sci.* USA, 74 : 560-564.

Mc Pherson, M.J., R.J. Oliver and S.J.Gurr. 1992. The polymerase chain reaction.   In : Molecular Plant Pathology : A practical approach Vol. I (Eds. Gurr, S.J.,

McPherson, M.j. and Bowless, D.J.), Oxford University, Press, 123-145.

Mendel, M. and Higa, A. 1970. Calcium dependent bacteriophage DNA detection. *J.Mol.Biol.* 53 : 159-62.

Mendoza, L.G., McQuary, P., Morgan, A., Gangadharan, R., Brignac, S. and Eggers, M. (1999). High throughput microarray-based enzyme-linked immunsorbent assay (ELISA). *Biotechniques* 27: 780-787.

Miller, H.J. (1983). Some factors influencing immunofluorescence microscopy as applied in diagnostic phytobacteriology with regard to Erwinia amylovora. *Phytopath. Z.* 108: 235-241.

Miller, H.J. (1984). A method for the detection of latent ringspot in potatoes by immunofluorescence microscopy. Potato Res.27: 33-42.

Miller, S.A. and Martin, R.R. (1988). Molecular diagnosis of plant disease. *Annu. Rev.Phytopath.* 26: 409-32.

Miller, S.A., Grothaus, G.D., Petersen, F.P., Rittenburg, J.H. and Lankow, R.K. (1987). Detection and monitoring of turfgrass pathogens by immunoassay. Extended abstract. Environ. Chem. Div. Am. Chem. Soc., New Orleans, LA, Aug, 30 Sep.4.

Milne, R.G 1993.   Electron microscopy of *in* vitro preparations.   In Diagnosis of Plant viruses (ed: Matthews, R.E.F.), CRC, Boca Raton, pp 215-251.

Milne, R.G. and Luisoni, E. (1975). Rapid high resolution immune electron microscopy of plant viruses. *Virology* 68: 270-274.

Miniatis, T., Fritsch E.F. and Sambrook J. 1982. Molecular cloning : a laboratory manual. Cold Spring Harbour Laboratory Press, New York.

Mitchell J P and Gildow F E 1975 The initiation and maintenance of *Vicia faba* tissue cultures. Physiologia Plantarum 34: 250-253.

Mitchell, L.A. (1986). Derivation of Sirococcus strobilinus with monoclonal antibodies in an enzyme-linked immunosorbent assay. Can. *J. For. Res.* 16: 945-48.

Mitter N, Sulistyowati E, Graham M W, Dietzgen R G, 2001. Supression of gene silencing: A threat to virus-resistant transgenic plants? Trends Plant Sci, 6: 246-247.

Mueller, W.C., Tessier, B.J. and Englander, L. (1986). Immuno cytochemical detection of fungi in the roots of Rhododendron. *Can. J. Bot.* 64: 718-23.

Mullis, K. (1987). Process for amplifying nucleic acid sequences. U.S. Patent No.4683202.

Mullis, K.B., F.Faloon, S.Scharf, R.Saiki, G.Horn and H.Erlich. 1986. Specific enzymatic amplification of DNA in vitro. The polymerase chain reaction. Cold spring Harber Symp. *Quart Biol.* 51 : 263.

Mumford, R.A., Barker, I. And Wood, K.R. (1996). An improved method for the detection of tospoviruses using the polymerase chain reaction. J. *Virol. Methods* 57: 109-115.

Murashige T 1961 Supression of shoot formation in cultured tobacco cells by gibberellic acid. *Science* 134: 280.

Murashige T 1974 Plant Propagation through tissue cultures. *Annual Review of Plant Physiology* 25: 135-166.

Murashige T and Skoog F 1962 A revised medium for rapid growth and bioassays with tobacco tissue cultures. *Physiologia Plantarum* 15: 473-497.

Murashige T and Tucker D P H 1969 Growth factors required for citrus tissue culture. In proceedings of Ist International Citrus symposium. Reverside, University of California 3: 1155-1161.

Narasimhan, M.J. (1968) on the presence of aflatoxins in copra. Hindustan *Antibiot. Bull* 2: 104-105 (22).

Nejidat A and Beachy R N, 1990. Transgenic tobacco plants expressing a coat protein gene of tobacco mosaic virus are resistant to some other tobamoviruses. *Mol Plant Microbe Interact*, 3, 247-251.

Nicolaisen, M., Rasmussen, H.N., Husted, K. and Nielsen, S.L. (2001). Reverse transcription detection of immobilized, amplified product in a one-phase system (RT-DIAPOPS) for the detection of potato virus Y. *Plant Pathol.* 50: 124-129.

Nie, X. and Singh, R.P. (2001). A novel usage of random primers for multiplex RT-PCR detection of virus and virioid in aphids, leaves and tubers. *J.Virol. Methods* 91:37-49.

Niepold, F. and Schober-Butin, B. (1995). Application of the PCR technique to detect Phytophthora infestans in potato tubers and leaves. *Microbiol. Res.*150: 379-385.

Norman W. Schaad; Reid D. Frederick, Joe show; William L. Schneider; Robert Hickson; Michael D. Petrillo and Douglas G.Luster,2003. Advance in molecular –based diagnostics in meeting crop biosecurity and Phytosanitary issues. *Annu Rev. Phytopathol.* 41: 305-24.

Owens, R.A. and Diener, T.O. (1981). Sensitive and rapid diagnosis of potato spindle tuber viroid by nucleic acid spot dhybridization. *Science* 213:67-672.

Palfreyman, J.W. (1998). Use of molecular methods for the detection and identification of wood decay fungi. Pp. 305-319. In: Forest Products Biotechnology (A Bruce, JW Palfreyman Eds.) Taylor and Francis, London.

Palukaitis P and Zaitlin M, 1997. Replicase-mediated resistance to plant virus disease. *Adv Virus Res*, **48**, 349-377.

Pappu H R, Niblett C L and Lee R F, 1995. Application of recombinant DNA technology to plant protection : Molecular approaches to engineering virus resistance in crop plants. *World J Microbial Biotechnol*, 1, 426-437.

Phillips, R.D. Hayes, A.W. and Berndt. W.O. (1980). High performance liquid chromatographic analysis of the mycotoxins citrin and its application to biological fluds. *J. Chromatography*. 190: 419-427. (23).

Pijut P M, Lineberger R D, Domir S C and Schreiber I R 1988 Response of elm callus to culture filtrate of *Ceratocystis ulmi* and correlation with whole plant disease reaction. *Horticultural Science* 23 (3): 787.

Podleckis, E.V.,Hammond, R.W., Hurtt, S.S. and Hadidi, A. (1993). Chemiluminescent detection of potato and prome fruit virioids by digoxigenin labeled dot blot and tissue blot hybridization. *J.Virol. Methods* 43: 147-158.

Pohland, A.E. and Wood, G.E. (1987). Occuance of mycotoxins in foods. Patges 35-64. In: Mycotoxins in food (Eds Krogh, P.) Academic Priss, London. (24).

Pooler, M.R, Myung, I.S., Bentz, J., Sherald, J. and Hartung, J. (1997). Detection of *Xylella fastidiosa* in potential insect vectors by immunomagnetic separation and nested polymerase chain reaction *Lett. Appl. Microbiol.* 25: 123-126.

Powell, P.A., Stark, D.M., Sanders, P.R and Beachy, R.N. 1989. Protection against tobacco mosaic virus in transgenic plants that express tobacco mosaic virus antisense RNA. *Proc. Natl.Acad. Sci.,* USA. 86: 6949-6952.

Powell-Abel P, Nelson R S, De B, Hoffman N, Rogers S G, Fraley R T and Beachy R N, 1986. Delay of disease development in transgenic plants that express the tobacco mosaic virus coat protein gene. *Science,* 232, 138-143.

Prasad, N.H. 2000. Role of Host-Genotypes of Cotton on plasmid Profile of Xanthomonas axanopodis pv.malvacearum races. Ph.D. Thesis, IARI, New Delhi. Qiagen Plasmid Mini/Midi Handbook, 2000.

Ramaiah, M., Sankaralinga A., Rajappan K., Rabindran R., Krishnaveni A., 2003 Biological, Seriological, Electron Microscopic, Tissue culture and Molecular technique for the early detection and elimination leading to management of plant viruses in agro ecosystem perspectives – Practical manual, Centre for plant protection studies, TNAU, Coimbottor.

Rainer Nicolay and Richard A. Sikora, 1988. The proved techniques for the detection of nematophagous fungi and their activity against target nematodes. Revue Nematol 11(i),: 115 – 116.

Raymond S and Weintraub L 1959 Acrylamide gels as a supporting medium for zone electrophoresis. *Science* 138: 711.

Reimann-Philipp U, 1998. Mechanisms of resistance: Expression of coat protein. *Methods Mol Biol,* 81, 521-532.

Reinert J 1959 Uber die kontrolle der Morphogenese and die induction von adventive embyonen and gewebekulkuremaus karotten. *Planta* 53: 318-338.

Rezain, M.A., Skene, K. G.M and Ellis, J.G. 1998. Antisense RNAs of cucumber mosaic virus in transgenic plants assessed for control of the virus. *Plant Mol Biol.* 11: 463-471.

Robbins W J 1922 Cultivation of excised roots tips and stem tips under sterile conditions. *Botanical Gazette* 73: 376-390.

Robert R.Martin; Delano James, and Andre Levesque.C 2000. Impacts of molecular diagnostic technologies of plant disease management. *Annu. Rev. Phytopathol.* 38; 207-39.

Robold, A.V. and Hardham, A.R. (1998). Production of species-specific monoclonal antibodies that react with surface components on zoospores and cysts of *Phytophthora nicotianae*. Can. J. Microbiol. 4: 1161-1170.

Rybicki, E.P. and VonWechmar, M.B. (1982). Enzyme assisted immune detection of plant virus proteins electroblotted onto nitrocellulose paper. *J.Virol. Methods*. 5: 267-278.

Sambrook, J, Fritsch, E.F. and Maniatis, T. 1989. Molecular cloning : A laboratory manual. Cold Spring Harbour Laboratory press, New York.

Sanford J C and Johnson S A, 1985. The concept of parasite-derived resistance genes from the parasite's own genome. J theor Biol, 113, 395-405.

Sanford J C and Johnston S A, 1985. The concept of parasite-derived resistance genes from the parasite's own genome. J Theor Biol, 113, 395-405.

Sanger, F., Nicklen, S. and Coulson, A.R. 1977. DNA sequencing with chain terminating inhibitors. *Proc. Natl. Acad. Sci.* USA, 74 :544463-5467.

Scala A, Bettini P, Bniatti M, Bogani P, Pellegrini G and Togoni F 1984 *In vitro* analysis of the tomato *Fusarium oxysporum* system and selection experiments. Plant-Tissue-Cell-Culture for Applied Crop Improvement (1984) meet : 361-362.

Schaad, N.W. (1978). Use of direct and indirect immunofluorescence tests for identification of Xanthomonas campestris. *Phytopathology* 68: 249-252.

Schaad, N.W. (1979). Serological identification of plant pathogenic bacteria. *Annu. Rev. Phytopath*. 17: 123-147.

Schaad, N.W. and Frederick, R.D., Shaw, J., Schenelder, W.L, Hickson, R, Petrillo, M.D. and Luster.D.G. (2003). Advances in moleculr-based diagnostics in meeting crops biosecurity and Phytosanitary issues. *Annu. Rev. Phytopath*. 41: 305-24.

Schaad, N.W., Berthier-Schaad, Y., Hatziloukas, E., and Knorr, D. (1997). Development and comparison of ABI Prism 7700 sequence detection system to BIO-PCR for sensitive detection of *Pseuodomonas syringae* pv. *phaseolicola*. ASM Conf. Mol. Diagnostics Ther., Kananaskis, Alberta, Can Publ. 40.

Schaad, N.W., Berthier-Schaad, Y., Sechler, A. and Knorr, D. (1999). Detection of *Clavibacter michiganensis* subsp. *sepedonicus* in potato tubers by BIO-PCR and automated real-time fluorescence detection system. *Plant Dis*. 83: 1095-1100.

Schaad, N.W., Hatziloukas, E. and Hanricksen, G. (1996). Development of membrane BIO-PCR for ultra-sensitive detection of pathogens in environmental samples. *Am.Soc. Microbiol. Annu.* Meet., (Abstr)p. 391.

Schaad, N.W., Jones, J.B., and Chun, W.(Eds.). (2001). Laboratory Guide for identification of Plant Pathogenic Bacteria. APS Press, St.Paul, MN.

Schaad, N.W., Opgenorth, D. and Gaush, P. (2002). Real-time PCR for one-hour on-site diagnosis of pierce's disease of grape in early season asymptomatic vines. *Phytopathology* 92: 721-28.

Schade, J.E. and king. Jr., A.D. (1984). Analysis of major Alternaria toxins. *J. Food Prot.* 47: 978-995 (25).

Schoen, C.D., Knorr, D. and Leone, G. (1996). Detection of Potato leafroll virus in dormant potato tubers by immunocapture and fluorogenic 5' nuclease RT-PCR assay. *Phytopathology* 86: 993-999.

Schubert, R, Bahnbeg G. Nechwatal J. Jung T, and Cooke, D.E.L. (1999). Detection and quantification of Phytophthora species which are associated with root rot diseases in European decidous forests by species-specific polymerase chin reaction. *Eur. J. for. Pathol.* 29: 169-188.

Sears, B.B. and KIRkpatrick, B.C. (1994). Univellling evolutionary relationships of plant pathogenic mycoplasmalike organisms. ASM News 60 : 307-12.

Seemuler, E. Kison, H. Lorenz, K.H.Schnieder, B., Marcome, C., Smart, C.D. and Kirkpatrick, B.C. (1998). Detection and identification of fruit tree Phytoplasmas by PCR amplification of ribosomal and mitochondrial DNA, pp.56-66 in: New Technologies for the improvement of plant Disease Diagnosis. (C.Manceau and JS Pak Eds.) off Publ. Eur. Common., Luxembourg.

Seoh, M.L,, Wong, S.M. and Zhang, L. (1998). Simultaneous TD/RT-PCR detection of cymbidium mosaic potexvirus and odontoglossum ringspot tobamovirus with a single pair of primers. *J. Virol. Methods* 72: 197-204.

Shepherd S L K 1986 Selection for early blight disease resistance in tomato : use of tissue culture with *Alternaria solani* culture filtrate. International Congress on Plant-Tissue-Cell-Culture 6th meeting 211.

Sherwood JL, 1994. On efforts to improve detection of selected groundnut viruses and expression of tomato spotted wilt virus antibodies in plants. in Working together on Groundnut Virus Disease, edited by DVR Reddy DVR, McDonald D & JP, ICRISAT, Hyderabad, India. Pp 48-48.

Sitari, H. and Kurppa, A. (1987). Time-resolved fluoroimmunoassay in the detection of plant viruses. *J. Gen. Virol.* 68: 1423-1428.

Simpson S F, Libenga K R and Torrey L G 1974 The effective auxin for cytodifferentiation in pea root cortical explants. *Plant Physiology* 11: 13-20.

Singh, A and Sinha K.K. (1982 a). Aflatoxin production on some fruits by Aspergillus flavus Link ex Fries and Aspergillus parasiticus Spere. Curr. Sci. 51 : 282 (27).

Singh, A and Sinha K.K. (1982b). Biochemical changes in musambi fruits inoculated with species of aflatoxin producting aspergilli. Curr. Sci. **51**: 841-842 (28).

Singh, A and Sinha K.K. (1983). Biochemical changes and aflatoxin production in guava fruits by *Aspergillus flavus* and *A.parasiticus*. *Indian Phytopath*. **36** : 365-366 (29).

Singh, A. (1983) Mycotoxin contamination in dry fruits and spices. Pages 55-68. In : Mycotoxins in Food and Feed. (Eds-K.S.Bligrami, T. Prasad, and K.K. Sinha Allied Press, Bhagalpur (26).

Singh, M. and Singh, R.P.(1995). Digoxigenin-labeled cDNA probes for the detection of potato virus Y in dormant potato tubers. *J. Virol. Methods* 52: 133-143.

Singh, M.N. and Khurana, S.M. Paul (2000). Role of additives in sample extraction buffer for improvement of ELISA for potato leaf role virus, pp. 439-443. In: Potato: Global Research and Development vol (Eds. Khurana, S.M. Paul, G.S. Shekhawat, B.P. Singh & S.K. Pandey), Indian Potato Association, Shimla.

Singh, M.N., Khurana S.M. Paul and Joshi, U.M. (1989). A simple penicillinase based ELISA for potato virus X. *Proc. Indian Nat. Sci. Acad*. 55(B): 287-290.

Singh, M.N., Mukerjee, K., Khurana, S.M.Paul, Gopal, Jai and Querci, M. (2000b). Detection of potato spindle tuber viroid by NASH in exotic potato germplasm. pp. 219-235 In: Potato: Global Research and Development vol (Eds. Khurana, S.M. Paul, G.S. Shekhawat, B.P. Singh & S.K. Pandey), Indian Potato Association, Shimla.

Singh, R.P. (2000). Advances in molecular detection methodologies for potato viruses and virioids. Pp.219-235. In: Potato : Global research and Development vol (eds. Khurana, S.M.Paul, G.S. Shekhawat, B.P. Singh & S.K. Pandey), Indian Potato Association, Shimla.

Singh, R.P., Boucher, A., Lakshman, D.K. and Taventzis, S.M.(1994). Multimeric non-radioactive cRNA probes improve detection of potato spindle tuber viroid (PSTVd). *J. Virol. Methods* 49: 221-234.

Singh, Sarjeet, Barker, H. and Kumar, Shiv (2000). Prospect of penicillinase ELISA for detection of plant viruses in the developing countries, pp. 491-494. In: Potatop: Global Research and Development vol (Eds. Khurana, S.M. Paul, G.S. Shekhawat, B.P. Singh & S.K. Pandey), Indian Potato Association, Shimla.

Sinha, K.K. and Singh. A. (1982a). Biochemical changes in orange fruits during infestation with *Aspergillus flavus* and A.parasiticus. Nat. Acad. *Sci. Letters* 5: 143-144. (30).

Skoog F and Miller C O 1957 Chemical regulation of growth and organ formation in plant tissues cultured *in vitro. Symposia Society for Experimental Biology* 11: 118-131.

Southern, E.M. 1975. Detection of specific sequences among DNA fragments separated by gel electrophoresis. *J.Mol. Biol.* 98 : 503-517.

Stead, D.E. (1999). Validation of diagnostic methods for diseases such as potato ring rot and potato brown rot for use within the European Union. In: Program Book, p.68 APS/CPS joint Meet, Montreal, Aug.7-11.

Stoloff, L. (1976). Occurrence of mycotoxins in foods and feeds. Adv., Chem. Ser. 149: 23-50(34).

Swaminathan, M.S. (2003) Enhancing our agricultural competitiveness. *Curr. Sci.* 85: 886-895.

Szemes, M., Klerks, .M.M., Van den Heuvel, J.F. and Schoen, C.D. (2002). Development of a multiplex amplified Det RNA assay for simultaneous detection and typing of potato virus Y isolates. *J. Virol. Methods* 100: 83-96.

Tabler M, Tsagris M and Hammond J, 1998. Antisense RNA and ribozyme-mediated resistance to plant viruses. in Plant Virus Disease Control edited by A Hadidi, R K Khetarpal & H Koganezawa, APS, Minnesota, USA. Pp 79-93.

Tavladoraki, P., E. Benvenuto, S. Trinca D. De Martinis, A. Cattaneo and P.Galeffi. 1993. Transgenic plants expressing a functional single chain FV antibody are specially protected from virus attack. *Nature Lond.* 366, 469-472.

Terrada, E., Kerschbaumer, R.J., Giunta, G., Galeffi, P., Himmler, G. and Cambra, M. (2000). Fully "recombinant enzyme-linked immunosorbent assays' using genetically engineered single-chain antibody fusion proteins for detection of Citrus tristeza virus. *Phytopathology* 90: 1337-1344.

Thomas P E, Hassan S, Kaniewski WK, Lawson E C & Zalewski J C, 1998. A search for evidence of virus/ transgene interactions in potatoes transformed with the potato leaf roll virus replicase and coat protein genes. *Mol Breed*, 4. 407-417.

Thomas, C.M., Vos, P., Sabeen, M., Jones, D.A., Narcoh, K.A., Chandwink, B.P. and jones, J.D.G. 1995. Identification of amplified restriction fragment polymorphism (AFLP) markers tightly linked to the tomato Cf-9 gene for resistance to *Cladosporium fulvam*.   *The plant Journal* **8** : 785-794.

Torrance, L. (1987). Use of enzyme amplification in an ELISA to increase sensitivity of detection of barley yellow dwarf virus on oats and individual vector aphids. *J.Virol. Methods* 15: 131-138.

Torrance, L. (1995). Use of monoclonal antibodies in plant pathology., *J. Plant Pathol.* 101. 351-363.

Torrance, L. and Jones, R.A.C. (1982). Increased sensitivity of detection of plant viruses obtained by a fluorogenic substrate in enzyme-linked immunosorbent assay. *Ann. Appl. Biol* 101: 501-509.

Toth, K.F., Harrison, N.A. and Sears, B.B. (1994). Phylogenic relationships among members of the class Molicutes deduced from rps3 gene sequences. *Int. J.Syst. Bacteriol* 44: 119-24.

Tricoli D M, Carney K J, Russell P E, McMaster J R, Geooff I D W, Hadden K C, Himmel P T, Hubbard JP, Boeshore ML & Quemada HD, 1995.   Field evaluation of transgenic squash containing single or multiple virus coat protein gene constructs for resistance to cucumber mosaic virus. Water*melon mosaic virus 2 and zucchini yellow mosaic virus.* Biotechnology, 13, 1458-1465.

Van Regenmortel, M.H.V., Fauguet, C.M., Bishop, D.H.L et al. (eds.). 2000. Virus Taxonomy: Classification and nomenclature of viruses. 7th report of International committee on Taxonomy on Viruses, Academic Press, San Diego.

Varma A, Jain R K. & Bhat I A, 2002. Virus resistant transgenic plants for environmentally safe management of viral diseases. *Ind J Biotech.*, 1: 73-86.

Varma, R.A.B. Prasad. J.S. and Verma. S.K. (1980). Aflatoxin production in fruits under market conditions. *J. Indian Bot. Soc.* **59**: 28 (35).

Vas, P., Hogers, R., Marjo, B., Reijon, M., Lee, T.Van de, Homes, M.,Frijters, A.,Pot, J., Peleman, J., Kuiper,M. and Zabeau, M. 1995. AFLP : a new technique for DNA fingerprinting. *Nucleic Acids Research,* **23** : 4407-4414.

Vasil V and Hilderbrandt A C 1965 Growth and tissue formation from single isolated cells in micro culture. *Science* 147: 1454-1455.

Venkatachalam P and Jaybalan N P 1995 In vitro screening of groundnut cell lines resistant to crude toxin filtrate of *Cercosporidium personatum.* *Advanced Modern Biotechnology* 3 (2): 31.

Verma, Yogita, Khurana, S.M. Paul, Mohan, Jitendra and Mukerjee, Krishanu (2004). Development of indigenous nucleic acid probe for the detection of potato virus Y. *Proc.Nat. Acad.* Sci. India 74(III & IV): 245-251.

Verma, Yogita, Shivali, S., Ahlawat, Y.S., Khurana, S.M.Paul, Nie, X. and Singh, R.P. (2003). Evaluation of multiplex reverse transcription polymerase chain reaction (RT-PCR) for simultaneous detection of potato viruses and strains. *Indian J. Biotech.* 2: 587-590.

Visconti. A. Logrieco. A. and Bottalico. A. (1986). Natural occurrence of Alternaria mycotoxins in olives their production and possible transfer into the oil. *Food Addit. Contesm.,* 3: 323-330 (36).

Wang P, Zoubenko O and Tumer N E, 1998. Reduced toxicity and broadspectrum resistance to viral and fungal infection in transgenic plants expressing pokeweed antiviral protein II. Plant Mol. Biol, 38, 957-964.

Ward, L.J. and Deboer, S.H. (1994). Specific detection of Erwinia caratovora sub sp. atroseptica with a digoxigenin labeled DNA probe. *Phytopathology* 84: 180-186.

Waterhouse P M, Wang M Bo & Lough T, 2001. Gene silencing as an adaptive defence against viruses. *Nature Lond,* 411, 834-842.

Watson, D.H. (1984). An assessment of food contamination by toxic products of Alternaria, *J. Food. Prot.* **47**: 485-488.

Weisburg, W.G., Tully, J.G., Rose, D.L., Petzel, J.P. and Oyaizu, H. (1989). A phylogenetic analysis of the mycoplasmas: basis for their classification. *J. Bacteriol.* 171: 6455-67.

White P R 1939 Potentially unlimited growth of excised plant callus. In an artificial medium. *American Journal of Botany* 26: 59-64.

White, T.J., Bruns, T., Lee, S. and Taylor, J. (1990). Amplification and direct sequencing of fungal ribosomal RNA genes for phylogenetics, pp. 315-322. In: PCR Protocols, A Guide to Methods and Applications (MA Innis, DH Gelfand, JJ Sninsky and TJ White Eds.). Academic Press, San Diego, USA.

Wilson, W.J., Widemann, M., Dillard H.R. and Batt, C.A. (1994). Identification of Erwinia stewartii by a ligase chain reaction assay. *Appl. Environ. Microbiol.* 60: 278-284.

Woese, C.R, Maniloff, J. and Zablen, L.B. (1980). Phylogenetic analysis of the mycoplasmas. *Proc. Natl. Acad. Sci.* USA 77: 494-98.

Woese, C.R. (1987). Bacterial evolution. *Microbiol. Rev.* 51:221-71.

Yespes, L.M., Fuchs, M., Slightom, J.L and Gonsalves, D. 1996. Sense and antisense coat protein constructs confer high levels of resistance to tomato ringspot nepovirus in transgenic Nicotiana species. *Phytopathology.* 86: 417-424.

Zaitlin M, Anderson JM, Perry KL, Zhang L & Palukaitis P, 1994. Specificity of replciase mediated resistance to cucumber mosaic virus. *Virology*, 201, 201-205.

www.ingramcontent.com/pod-product-compliance
Lightning Source LLC
Chambersburg PA
CBHW031949180326
41458CB00006B/1669